Guide to
RIBA Domestic and Concise Building Contracts 2018

RIBA ⬛ Publishing

Sarah Lupton

Guide to RIBA Domestic and Concise Building Contracts 2018

© Sarah Lupton, 2018

Published by RIBA Publishing,
66 Portland Place, London, W1B 1AD

ISBN 978 1 85946 863 0

British Library Cataloguing-in-Publication Data
A catalogue record for this book is available from the British Library.

Commissioning Editor: Alexander White
Project Editor: Daniel Culver
Production: Jane Rogers
Designed and Typeset by Academic + Technical, Bristol, UK
Printed and bound by W&G Baird Ltd, Great Britain
Cover design: Kneath Associates
Cover Image credit: © David Papazian / Shutterstock.com

www.ribapublishing.com

Foreword

In a rapidly changing world, it is reassuring to know that the RIBA continues to provide stability, simplicity and common sense with straightforward contracts for architects, clients and contractors alike. Offering security and fairness in equal measure, the RIBA Domestic and Concise Building Contracts have been updated in 2018 to respond directly to user needs and the result is an accessible and easy-to-use set of contracts for all parties. Nevertheless, contracts can always prove challenging for those with little experience and little time to spare.

Sarah Lupton's guide is the perfect companion for navigating the new contracts. The handy walkthroughs, advice and tips will give readers the confidence to prepare and finalise the contracts with all parties understanding the implications and their roles. The guide is very accessible, with comprehensible language suitable for newcomers and experienced users alike. A number of features have been added, including a new section on practical completion, to give the guide additional depth and value for the user. The expanded chapters offer even more practical information on essential matters such as choice of form, tendering and contract formation. Further, the author's extensive experience is particularly invaluable when it comes to presenting challenging topics such as liabilities and termination in a clear, well-structured manner.

The success of these evergreen contracts for smaller works and the practical relationships they describe makes them lastingly popular. All members of the team should be given a copy of this guide so they are familiar with the domestic and concise forms of contracts, and can understand how they can be used for any scale of project. Indeed, the success of these contracts thus far has led to demand for their use in larger projects, in circumstances beyond those that were initially envisaged.

I congratulate the team for these most recent updates and I am proud as President of the RIBA to endorse their use by members and others to deliver excellence in service to our clients.

Ben Derbyshire
President of the RIBA

About the author

Professor Sarah Lupton MA, DipArch, LLM, RIBA, CArb is dual qualified in both architecture and law. She is a partner in Lupton Stellakis architects, and holds a personal chair at Cardiff University. As a specialist in design liability and construction law she also practises as a Chartered Arbitrator, adjudicator and expert witness, and is the author of a number of books, including *Which Contract?* (RIBA Publishing), the highly popular guides to the JCT contracts (RIBA Publishing) and *Cornes and Lupton's Design Liability in the Construction Industry* (Wiley-Blackwell). Sarah is the CIC Liability Champion, and chair of its Liability Panel.

Contents

About this Guide

The RIBA Concise Building Contract and RIBA Domestic Building Contract were first published in 2014, and were updated and streamlined in 2018 to take account of user feedback. The contracts have proved popular for use on small and medium-scale projects that follow a traditional procurement route, providing useful and innovative alternatives to other standard contracts available. They were designed to be flexible and suited to the needs of consultants and their clients, and the changes have helped to improve their clarity and ease of use.

This Guide assumes no prior knowledge of the contracts, or of construction contract law. As well as examining the provisions of the contracts, it also covers related legal topics, such as relevant legislation, contract formation and dispute resolution. There have been no reported cases involving the RIBA Building Contracts so, as in the previous edition of this Guide, the provisions have been analysed by analogy with existing case law on similar contracts. Many relevant new cases had been added to this edition of the Guide.

The first chapter of the Guide discusses the situations where it may be appropriate to use these contracts and outlines the contracts' key provisions. Detailed tables compare the contracts with each other and with other standard building contracts. The second chapter examines information that should be included in the tender package, and the formation and execution of the contract. The Guide then introduces the roles of the client, the contractor and the contract administrator, presenting tables of their duties and powers. The remaining chapters look in detail at programming and extensions of time, control of the works, certification and payment, insurance, termination and dispute resolution.

A new chapter has been introduced in this edition, focusing on issues related to practical completion, as this is an aspect of contract administration that can prove problematic in practice. An outline of relevant legislation such as the Housing Grants, Construction and Regeneration Act 1996 and the Consumer Rights Act 2015 is included in a new Appendix.

The contracts have retained a large number of the useful provisions that were in the 2014, such as the collaborative working provisions (pre-start meeting, advance warnings, joint resolution of delay, proposals for improvements and cost savings) and the optional provisions, for example for milestone payments, for a contractor programme, for required specialists (for whom the contractor takes entire responsibly), for contractor design, for new building warranties, and for Public Sector clients. All of these are discussed in the Guide.

In summary, the RIBA Building Contracts continue to offer an attractive option in that they are suitable for a wide range of projects, are relatively short and easy to read, and yet contain a range of innovative features not found in other standard forms.

1 Introduction

1.1 In November 2014 the Royal Institute of British Architects (RIBA) launched two new building contracts: The RIBA Concise Building Contract 2014 and the RIBA Domestic Building Contract 2014. The contracts featured in this book are the second editions of these contracts, published in January 2018.

1.2 The 2014 forms were not the first contracts to be published by the RIBA. In fact, the very first UK standard form of building contract was published by the RIBA jointly with the Institute of Builders and the National Federation of Building Trades Employers of Great Britain and Northern Ireland in 1902 (and cost one shilling!). In 1931 the Joint Contracts Tribunal (JCT) was formed, and the drafting and publishing of the contract became a pan-industry project. It continued to be called the RIBA Standard Conditions of Contract until 1963, when it was republished as the JCT Standard Form of Building Contract.

1.3 The JCT now publishes an extensive suite of contracts. However, the RIBA had identified a need, through feedback from its members, for short, easy-to-read and flexible contracts that would be suitable for less straightforward projects than those catered for by the JCT Home Owner contracts and that could be used on domestic and commercial works. It therefore published two new forms, designed for use in conjunction with the RIBA's architect/consultant appointment agreements.

1.4 The 2018 editions were revised to respond to user feedback, and to changes in legislation and industry practice. Although essentially the same forms, there are a significant number of changes, many of which are more than simply tidying up or minor drafting adjustments; in fact, most clauses have been revised to some extent. The most significant changes are outlined at paras. 1.27 and 1.28 below. The new forms are fully compatible with the new suite of RIBA Professional Services Contracts, also published in 2018.

Features of the RIBA Building Contracts

1.5 According to its guidance notes, the RIBA Concise Building Contract 2018 (CBC) is intended for use on 'all types of simple commercial building work'. It can be used in both the private and public sectors, as it includes optional provisions dealing with official secrets, transparency, discrimination and bribery as normally required by public sector clients (item Y and cl. 24). The RIBA Domestic Building Contract 2018 (DBC), as its name suggests, is intended for domestic work, including renovations, extensions, maintenance and new buildings. Its guidance notes state that this is limited to 'work carried out on the Client's home', i.e. to contracts that fall under the 'residential occupier exception' in the Housing Grants, Construction and Regeneration Act 1996, as amended by Part 8 of the Local Democracy, Economic Development and Construction Act 2009 (hereafter referred to as the Housing Grants Act) (see para. 1.21). DBC is endorsed by the HomeOwners Alliance.

1.6   Both contracts are available in hard copy and digital format (using the RIBA Contracts Tool), and can be purchased from: www.ribacontracts.com.

1.7   Key features of the RIBA Building Contracts are:

- collaboration provisions: pre-start meeting, advance warnings, joint resolution of delay, proposals for improvements and costs savings;

- flexible payment options;

- provision for contractor design;

- optional provisions for a contractor programme;

- optional provisions for client-selected suppliers and subcontractors;

- mechanisms for dealing with changes to the project which allow for agreement and include specified timescales;

- option for commencement and completion in stages;

- terms compliant with the Consumer Rights Act 2015 for consumer clients;

- inclusion of guidance notes on use and completion, together with a user checklist;

- client cancellation form included in DBC.

1.8   All of these features are discussed at various points throughout this Guide.

Suitability for different procurement routes

1.9   Both of the RIBA Building Contracts are suitable for projects that are procured on what is normally referred to as a 'traditional' procurement route, (or 'design-bid-build'), i.e. one where the client engages at least one, and possibly several, firms of consultants to prepare a design and complete full technical documentation before the project is tendered to contractors. Traditional procurement is widely used – in the 2018 NBS survey,[1] 46 per cent of respondents stated it was the method used most often on their projects (the equivalent figure for design and build procurement was 41 per cent).

1.10   With traditional procurement, it is common practice that some parts of the design are completed by the contractor or by specialist subcontractors (this version is sometimes referred to as 'traditional plus contractor design'). The client, through its consultant team, remains responsible for providing the majority of the design, and for overall design integration, but limited and carefully delineated sections are delegated to the 'supply chain'. The RIBA Building Contracts would be suitable in this situation, with either the main contractor or a subcontractor (client selected, if required) undertaking the design.

1.11   The RIBA Building Contracts are not intended for use in design and build procurement, i.e. where the contractor is responsible for both the design and the construction of the project. It would be possible to adapt them for this use, but there would still be a need for a contract document giving full details of the design requirements, and for an appointed

---

[1]   NBS National Construction Contracts and Law Report 2018.

contract administrator (not usual in design and build contracts). In addition, all the provisions relating to design would require amendment, and more detailed provisions regarding submission and approval of the developing design would have to be added.

1.12   The contracts are also not intended to be used with management contracting or construction management arrangements (where the project is tendered as a series of packages to separate firms, with work progressing on a rolling programme basis after the first package is let). Management procurement arrangements are normally used on very large projects; however, on a smaller scale, clients who wish to manage projects themselves sometimes adopt a similar system and engage a number of separate companies (often referred to as 'separate trades'). The RIBA Building Contracts might be suitable for some of the larger work packages, but thought should be given as to how all the separate contracts are to be coordinated. This is not an easy task, and the apparent savings achieved by cutting out the main contractor's mark-up may be more than offset by the amount of time the client has to spend managing the process, or by the additional fees charged by consultants if they undertake this role.

1.13   However, within the traditional plus contractor design route, the RIBA Building Contracts would be suitable for a wide range of projects, from very small-scale alterations and refurbishment works, to moderately sized projects relating to existing or new buildings. As mentioned above, CBC's guidance notes refer to 'simple' work, and generally it is the level of complexity, rather than the value, that should be the key determinant in the choice of contract.

1.14   Although the RIBA Building Contracts have many useful features (see para. 1.7), which mean they are flexible, some of the provisions lack the detail to be found in larger contracts. Examples are those relating to design submission and approval procedures and insurance clauses. Other provisions commonly found in larger contracts are not included, such as fluctuations clauses and contractor bonds – if these are required then other standard contracts should be considered.

Differences between the concise and domestic contracts

1.15   As a result of the redrafting in the 2018 editions, several of the differences between the two versions have been removed, and they are now much closer in layout and content. Immediately obvious is that all the idiosyncratic (and in many cases unnecessary) differences between clause numbering and wording have been removed. In addition, some of the unique provisions in each contract have been removed (the effect of these changes is discussed further below). There are still, however, a number of important differences, largely reflecting the intended use of DBC with consumer clients. A list of the differences between the two versions is set out in Table 1.1.

1.16   DBC has an option whereby the client may elect to act as contract administrator (cl. 19). It also has a provision allowing the client to cancel the contract within 14 days (cl. 12.11), and includes a cancellation form for this purpose.

1.17   A key difference relates to the dispute resolution procedures: in DBC adjudication is an optional provision, whereas in CBC adjudication is stated to be a contractual right (ensuring compliance with the Housing Grants Act, see Appendix 1).

Table 1.1 Key differences between CBC and DBC			
Title/subject	CBC clause	DBC clause	Notes
Agreement			Note on cancellation in DBC
Contract Details			
Client details	A	A	No reference to registered company in DBC
Dispute resolution	N	N	Differences in guidance
Client acting as contract administrator		T	DBC only
Advanced payment	T		CBC only
Collateral warranty/third-party rights	X		CBC only
Public sector clauses	Y		CBC only
Contract Conditions			
Client's cancellation rights		12.11	DBC only
Arbitration	13.8	13.4	Differences
Litigation	13.9	13.5	Differences
Advanced payment	19		CBC only
Client acting as contract administrator		19	DBC only
Collateral warranty/third-party rights	23		CBC only
Public sector clauses	24		CBC only

1.18   CBC includes optional clauses that provide for advance payment. It also includes an optional clause dealing with collateral warranties and third-party rights (see para. 2.29). These are usually provided to future purchasers or tenants, or to funders. The DBC does not provide for collateral warranties as such warranties do not normally arise in a domestic context. In the unlikely event that the client's bank requires a warranty, then an additional provision would need to be introduced. Finally, CBC includes public sector clauses, which will enable it to be used by public bodies such as local authorities.

1.19   Some of the differences between the forms arise from legislation, and its impact on contracts in the intended context for DBC and CBC. (The implications are outlined briefly below, and the legislation discussed in more detail in Appendix 1.)

Use by domestic and commercial clients

1.20   DBC is obviously intended for domestic projects, and CBC for commercial ones. However, identifying the precise differences between different categories of users is not quite as simple as it might appear – this stems from the different definitions used in legislation that may apply to construction contracts and/or contracts with consumers.

1.21   A key issue in such cases is whether the client is a 'residential occupier' for the purposes of the Housing Grants Act 1996. If the client is not a 'residential occupier', then the Housing Grants Act 1996 provisions regarding payment (including payment and pay less notices) and the right to adjudication are required by law and, conversely, they are not needed where the client is a residential occupier. The RIBA has made adjudication an optional provision in DBC, but has included all the Housing Grants Act payment requirements in both versions of the contract. Some thought should be given as to whether a residential occupier client would wish to include these provisions or would prefer a simpler payment regime, in which the requirement to issue pay less notices, etc. could be omitted.

1.22   A further distinction is whether the client is a 'consumer' for the purposes of the Consumer Rights Act 2015 and the Consumer Contracts (Information, Cancellation and Additional Charges) Regulations 2013.

1.23   DBC includes the right of the client to cancel the contract within 14 days of signing it (cl. 12.11), something required by the Consumer Contracts (Information, Cancellation and Additional Charges) Regulations 2013. This right is not included in CBC. Where CBC is being used with a consumer client, it might be wise to add this provision, or to make it clear that the client would in any case have this right under the Regulations.

1.24   The law affords special protection to persons who enter into a contract as a 'consumer', through what is referred to collectively as consumer protection legislation. A key Act is the Consumer Rights Act 2015, which implies particular terms into contracts for work and materials, and for services, and also will render void any terms that may be considered 'unfair'. A term will be considered unfair if it is not clear (among other things) and if it significantly affects the balance of rights between the parties. Both of the RIBA Building Contracts have been drafted with the intention that they should be fair. The language is certainly clear and should be relatively easy for a lay client to understand, and the key agreement regarding subject matter and price are clear in the articles and contract particulars. It should be noted that in *Domsalla* v *Dyason* the court stated that it considered the withholding notice terms to be unfair, however this was in relation to the Unfair Terms in Consumer Contracts Regulations 1999, which no longer apply, and in a case with very unusual facts. It is very unlikely that in normal circumstances the equivalent terms in the RIBA forms would be considered unfair.

> **Domsalla v Dyason [2007] BLR 348**
>
> The employer, Mr Dyason, whose house had burnt down, was advised by his insurers to enter into a contract based on MW98 with the contractor, Domsalla. The employer saw the form for the first time at the first site meeting, when he was presented with it for signature. The project was delayed, and the employer did not pay the sums certified in the last three certificates for payment, totalling £127,871.33, but no withholding notices were served. The contractor initiated an adjudication and the adjudicator decided in its favour against the owner. Subsequently, the Technology and Construction Court refused to enforce the decision and the employer was given leave to defend the claim. Part of the grounds for this was that the withholding notice provisions of the contract were considered unfair (under the Unfair Terms in Consumer Contracts Regulations 1999) in the special circumstances of the case, for example that the client was unaware that the terms existed, and that all notices were to be issued by the insurers.

1.25   It is important to note that the definitions of 'residential occupier' and 'consumer' are not the same. The usual definition of consumer in relevant legislation is a person who is acting 'for purposes which are outside his trade, business or profession'. This could include, for example, a wealthy business tycoon who is developing a country estate with several large residences for members of his family; in other words, it is not confined to small projects, or to those undertaking work on their own home, or to residential occupiers. Therefore, although a residential occupier will always be a consumer (i.e. the client in DBC will be a consumer), there may also be situations where a consumer is not a residential occupier, in which case CBC would be the appropriate choice.

Use by public bodies

1.26   CBC includes provisions relevant to any client that is subject to the Freedom of Information Act 2000 (which would include local authorities). These provide that the client is responsible for disclosing information as required by that Act, and that the contractor shall pass all requests for information to the client (cl. 24.2). The client has the discretion to determine what material may be exempt (cl. 24.3). Clause 24.4 prohibits the contractor from practising or allowing any form of discrimination, corrupt practice or bribery in the carrying out of the works.

Changes to the forms since the previous editions

1.27   There are a considerable number of changes since the previous editions, and as noted above the majority of clauses have been revised to a degree. The majority of these changes are to tidy up the drafting and regularise the clause numbering, but there are in addition quite a number of changes that significantly affect the content of the contracts. The key changes are listed below:

- checklists for completion added;
- several defined terms added;
- requirements for progress meetings removed;
- provisions regarding confirmation of verbal instructions removed;
- interim and final certification and payment provisions regularised so that they are the same in both contracts (e.g. pay less notices added to DBC);
- time bar provisions (i.e. provisions whereby the contractor loses a right if it fails to apply within a specified period) removed;
- detailed requirements for the optional programme, and sanctions for non-production removed;
- 'fit for purpose' option in relation to contractor design removed;
- client cancellation form added to DBC.

1.28   The tidying up is of course to be welcomed, and the contracts are generally clearer and better organised. The regularising between the two forms is, however, a mixed blessing. It makes them easier to use, particularly if a practice is likely to use both – all similar clauses are now identical, and where there are differences these are clear. However, in doing so the DBC now contains some of the Housing Grants Act payment requirements, in particular the need to issue pay less notices in advance, which could be onerous on a consumer client. It is also a pity to see some of the more innovative features removed, for example the time bar clauses and the sanctions for non-production of a programme. These provisions tipped the balance of risk towards the contractor. The balance in the 2018 versions is now closer to that in other contracts, such as JCT forms, although still more protective of the client than those contracts. (It is understood that the working group on the new edition, which included client representation, decided that the balance of risk should be adjusted to make the form more acceptable to all parties.) Nevertheless, many of the special features included in the 2014 editions have remained and the RIBA contracts retain a distinct combination of clarity and flexibility.

Comparison with other contracts

1.29   The RIBA Building Contracts are significantly different to others that are currently available. One striking feature is their length; they are shorter than most other contracts available for the intended scale of work, with the exception of the JCT Home Owner contracts, and possibly the NEC4 Engineering and Construction Short Contract (NEC4). However, they offer considerably more features, including a large number of optional clauses, than are available in competing contracts (the NEC4 Short Contract does not have the optional clauses that are contained in the full version). In fact, the RIBA contracts even offer some features that are not available in much larger contracts, as can be seen in Table 1.2.

Table 1.2 Comparison with other building contracts

	JCT Standard Building Contract (SBC)	JCT Intermediate Building Contract with contractor's design (ICD)	JCT Minor Works Contract with contractor's design (MWD)	JCT Building Contract and Consultancy Agreement for a Home Owner/Occupier	NEC4 Engineering and Construction Short Contract	RIBA Concise Building Contract 2018	RIBA Domestic Building Contract 2018
Contractor design	✓	✓	✓		✓	option	option
Client selected subcontractors	not for design	✓				option	option
CDP submission	✓	✓					
Professional Indemnity Insurance	✓	✓				✓	✓
Programme	✓				option	option	option
Sectional completion	✓	✓				✓	✓
Partial possession	✓	✓				✓	✓
Listed items	✓	✓					
Variation quotation	✓				✓	✓	✓

Table 1.2 Continued

	JCT Standard Building Contract (SBC)	JCT Intermediate Building Contract with contractor's design (ICD)	JCT Minor Works Contract with contractor's design (MWD)	JCT Building Contract and Consultancy Agreement for a Home Owner/Occupier	NEC4 Engineering and Construction Short Contract	RIBA Concise Building Contract 2018	RIBA Domestic Building Contract 2018
Loss/expense	✓	✓	limited	limited	✓	✓	✓
Variation and EOT rules	variation						
Client cancellation							✓
Insurance-backed guarantee						option	option
Risks register						referenced	referenced

1.30   In summary, the RIBA Building Contracts offer attractive alternatives to existing contracts in that they are relatively short and easy to read, and yet contain a range of innovative features not found in other contracts. This will make them appealing both to architects and their clients and to contractors.

2 Forming the contract

2.1   As noted in the previous chapter, the RIBA Building Contracts are most likely to be used on projects that follow a traditional procurement route, with possibly a degree of contractor and/or subcontractor design input. The decision to use a RIBA Building Contract will normally be made at an early stage in the project, and certainly by the end of the RIBA Plan of Work Stage 3,[1] as it affects issues such as the distribution of design responsibility between consultants and the contractor and the degree to which the client can control which subcontractors may be used.

2.2   Between the selection of the contract and the start of work on site, various processes will need to be undertaken: preparing the tender package and tendering; post-tender negotiations and appointment of the contractor; and formal execution of the contract. There will also be a pre-contract meeting, at which various matters are agreed. This chapter outlines these processes and considers particular matters that arise in relation to the RIBA Building Contracts.

Tendering

2.3   It can't be overemphasised that, when a project is sent out to tender, the contractor should be given full and detailed information regarding the project requirements. This is the principal means by which a designer can ensure that the quality of detailing, finish and workmanship will reach the required standards. If the project is not to be fully designed by the client's consultants, the contractor will require full information about the design that it is to provide, including any performance specifications.

2.4   Most importantly, full details should be given about which standard building contract is to be used, the particular conditions to be applied and any special terms. Consultants tend to focus on the design, technical and budget aspects, but clarity on the terms of the contract is just as important to the tenderer as these will affect the tender price. It is not advisable to introduce matters such as special contract provisions after a long tendering period, or after negotiations on price have been concluded: if they are not acceptable to the contractor, it will be in a very strong bargaining position. Furthermore, if contract documents are never formally executed, the details included in the tender package may subsequently form the basis of the contract between the parties (see para. 2.33).

Implications of the Consumer Rights Act 2015

2.5   This Act applies to contracts and notices between a 'trader' and a 'consumer'. A 'consumer' is defined as 'an individual acting for purposes that are wholly or mainly outside that individual's trade, business, craft or profession' (section 2(3)). For tendering purposes, it

---

[1]   The RIBA Plan of Work 2013 sets out key stages in a building project; an overview of the Plan of Work can be downloaded at: www.ribaplanofwork.com

should be noted that it gives consumers a certain amount of protection with regard to discussion that might be held during the tendering period. Essentially, any statements or promises that are made by the tenderer to the consumer prior to the contract being entered into, and which the consumer relies on, may be treated by the consumer as a binding term of the contract, whether or not they are ultimately embodied in the contract. Therefore, a great deal of care needs to be taken to ensure that any promises made during or after the tender are recorded and included in the eventual contract, or alternatively, if withdrawn by the tenderer, that the parties agree that they will not be included. Ideally the client should not enter into detailed discussions with separate tenderers – this should all be handled through their team of advisers.

Tendering procedures

2.6   Normally, for projects of the scale for which the RIBA Building Contracts are intended, one of two methods will be used: competitive tendering with a small number of contractors, or negotiation with a single contractor. The client's consultants will normally suggest which contractors should be considered, but the client may have worked successfully with firms before, or may have been recommended a firm by others, in which case those firms can also be included.

2.7   With competitive tendering, all tenderers should, of course, be sent identical information, so that they are competing on an equal basis. The tenders are then examined by the client's consultants, and normally the lowest priced one is accepted. With negotiated tendering, only the identified contractor will submit a tender, which is usually subject to discussion before acceptance.

2.8   An alternative method is a two-stage process: initially, a small number of contractors are asked to tender on the basis of less than complete information. Negotiations will then take place with the one that makes the most attractive submission. This method is frequently used where the contractor is expected to have a significant design input but the client wishes to approve this design before entering into the main contract; the selected contractor can then work with the client's consultants in finalising the design. It is unlikely that a two-stage tender would be used for a project based on one of the RIBA Building Contracts but, if it is, an agreement will be needed to cover the contractor's liability to the client for its contribution to the design process.

2.9   Full guidance on tendering procedures can be found in the *NBS Guide to Tendering: For Construction Projects* (Finch, 2011), JCT's *Tendering Practice Note 2017* (JCT, 2017) and the *RIBA Job Book* (Ostime, 2013), but the most important aspect for the purposes of this chapter is what to include in the tender documents.

Information that must be included in tender documents

2.10   The information included in the tender documents will eventually form the basis of the contract itself. Therefore, clearly, the tenderers must be told which version of the contract will be used, i.e. CBC or DBC, and whether it will be entered into as a deed or as a simple contract. It must also be clear which of the tender documents are to be priced by the contractor and, if the documents are to have a priority, what that priority is (see para. 2.16 below).

2.11   The identity of the client (including whether it is acting as an individual or a company) must be crystal clear, as should the nature of the works, the location of the site and any restrictions on its use and access. In fact, all matters that will later be entered in the 'Contract Details' (located at the front of the two contracts) should be set out in the tender documents. A full list of the items included in the Contract Details is given in Table 2.1.

CBC	DBC	Subject
Main items		
A	A	Client: name and details
B	B	Contractor: name and details
C	C	Description of the works; site address, and whether it will be occupied
D	D	Contract documents
E	E	Contract period: start date, date for completion and working hour restrictions
F	F	Facilities: permission for the contractor to use facilities
G	G	Architect/contract administrator
H	H	Other appointments
I	I	Consents, fees and charges; is the client or the contractor responsible for these?
J	J	Insurance
K	K	Contract price and payment frequency; rate of interest (if none given, will be statutory rate)
L	L	Liquidated damages (amount per day)
M	M	Defects fixing period (minimum 3 months)
N	N	Methods of dispute resolution
Optional items		
O	O	Programme: if one is required, and the details
P	P	Contractor design and Professional Indemnity Insurance (PII)
Q	Q	Required specialists
R	R	Completion in sections
S1	S1	Milestone payments
S2	S2	Payment on practical completion of the works
T	*not included*	Advanced payment
not included	T	Client acting as contract administrator
U	U	Insurance backed guarantee
V	V	Whether a new building warranty is required
W	W	Evidence of ability to pay the contract price
X	*not included*	Collateral warranties and third party rights
Y	*not included*	Public sector clauses

Table 2.1 Contract Details

2.12   Whoever is responsible for arranging the contract (usually the contract administrator) should discuss and agree all of this information with the client prior to tender. Some items are particularly important to raise at an early stage, so that the client has time to consider the matter and take advice, for example insurance and dispute resolution. Most are discussed in detail in other sections of this Guide (see the references in Table 2.1), but a brief introduction to key items in CBC and DBC is given here. There is also helpful information in the guidance notes to the contracts.

Whether the contract will be executed as a deed (Agreement: both contracts)

2.13   The decision as to whether the contract will be executed as a deed or as a simple contract (sometimes referred to as 'under hand') will affect the length of time the contractor will be held liable for a breach of the contract. Under the Limitation Act 1980, in the case of a simple contract the time limit for bringing an action for any breach is six years from the date of the breach, and in the case of a contract entered into as a deed it is 12 years. In the latter, the contractor's risk is obviously greater, which may result in higher tender prices. If no entry is made, the default will be six years.

The site, and whether the building will be occupied (item C: both contracts)

2.14   If the building is to be occupied during construction, this will have significant implications for the contractor's programme and the contract price. It is almost always slower, and more expensive, to work around partial occupation than it is if given a free run of the site. For example, added health, safety and security measures will be needed, such as the provision of protected access routes to the occupied parts. It will not be sufficient to simply indicate 'yes' as required: full details of the planned occupation will be needed (see para. 4.2). As well as indicating occupation, the client is also required to indicate any working hour restrictions (item E), and which facilities (electricity, parking, etc.) the contractor may use free of charge (item F, cl. 2.5).

Contract documents (item D, both contracts)

2.15   All the documents that will form the contract bundle should be clearly identified in the tender package. Many of these will have been prepared by the contract administrator and other consultants identified above, however there are some exceptions. If a contractor's design proposal is required, this will also be a contract document, and details of what is required from the contractor at tender stage, and possibly post-tender but before the contract is executed, should be set out (this should be referenced here and also under item P, see para. 2.27 below). The tender documents may include documents prepared by a required specialist, or by other consultants who are not listed.

2.16   In the 2014 edition the parties were invited to rank the documents in order of priority. Although this was optional, it was a sensible provision, and the parties could consider adding in a ranking, provided the ranking is carefully thought through and practical (feedback on the 2014 edition suggested it was not being applied properly). Ranking helps to avoid arguments later when, as is almost inevitable on most projects, it is discovered that there are inconsistencies between them (note that it would still be possible for a lower priority document to supplement the material in a higher priority document, however it would not be able to contradict it: see *RWE Npower Renewables* v *JN Bentley Ltd*). The two matters to consider are, of course, quality and quantity, and it may be that the ranking is different for both these, e.g. the specification being the key document for defining quality, and the bill of quantities for amount. It will be particularly important, where the contractor is undertaking design, to state whether the client's requirements for the design (probably included in the specification) or the contractor's design proposals prevail. Effectively this will determine whether, in cases of conflict, the contractor's primary obligation is to meet the specification requirements, or merely to provide what was set out in the contractor's proposals, even if it is later discovered that those proposals do not meet the client's design requirements. Normally a client would prefer the former approach.

The contract administrator and other appointments (items G and H, both contracts)

2.17 The names of the firms engaged by the client should be set out, rather than the individuals, although the main point of contact within each firm could be noted. The contract administrator has a key role in the contract (see Chapter 3) and, although identified, is not a party to the contract (cl. 5.1). The client may replace the contract administrator with another firm at any time, provided it notifies the contractor (cl. 1.3). The contract administrator may not be the only appointment; the client may also have engaged engineers, quantity surveyors and, possibly, a project manager. These should be listed under item H, and the client must inform the contractor of any later appointments as set out in cl. 1.3). The other consultants have no defined role under the building contract, and all communication should be with the contract administrator, but it is sometimes useful for the contractor to be aware of who else is advising the client.

Regulatory consents, fees and charges (item I, both contracts)

2.18 This item asks the parties to indicate who will be responsible for obtaining and paying for 'regulatory and statutory consents, fees and charges'. It will almost always be the case that the client takes responsibility for planning permission because, unless the project is to be entirely designed by the contractor, which is unlikely, the contractor will not be engaged at the time planning permission is usually sought. The application and any permission issues are normally resolved before going out to tender (the RIBA Plan of Work 2013 notes that 'Planning applications are typically made using the Stage 3 output'). It is possible, however, that the contractor may be required to deal with Building Regulations approval on small projects (by means of a building notice), but for more complex schemes the client will take responsibility. In the case of party wall consents, under the Party Wall etc. Act 1996 the building owner is required to serve notices on all adjoining owners of intended work on or near the site boundaries, and in many cases to appoint surveyors to act on its behalf. It is not possible to require the contractor to take on these roles under the building contract; a separate appointment would be needed (with either the contractor or another consultant).

Contract price (item K, both contracts)

2.19 Both contracts give two options for setting the contract price: 'Contract Price' or 'Contract Price calculated in accordance with the Pricing Document listed under item D'. It is not entirely clear what this distinction means. Even if the 'Contract Price' option is selected, which the guidance notes describe as the 'lump sum' option, the contractor would still be entitled to additional amounts in the event of changes to the works, etc. (see para. 6.6). A pricing document would be very useful to assess these changes, for example the pricing document could be a fully priced bill of quantities, or a schedule of works or a contractor-prepared price breakdown; therefore, selecting 'Contract Price' does not preclude the need for a pricing document.

2.20   The second option has no space for entering a total Contract Price, so the effect of selecting this option would to create a measurement, rather than a lump sum contract. What is probably intended is that for this option the pricing document will be in the form of a schedule of rates. This will give rates for all types of work envisaged, but not quantities, and will have been prepared by one of the client's consultants. Alternatively, if preferred, the contractor can be asked to prepare the document. The schedule of rates option is useful for projects where it is difficult to assess the likely quantity of work in advance, such as in refurbishment work, where the extent of repair work cannot be ascertained until demolition work is complete. However, there will be no overall contract sum, and it may be difficult to control the overall budget.

Liquidated damages (item L, both contracts)

2.21   The Contract Details require the parties to insert an agreed amount of liquidated damages per day, and the contractor should be informed of the rate at the tender stage. This is an amount to be allowed by the contractor in the event of failure to complete by the date for completion (cl. 10.1; note that if the works are divided into sections under optional clause 17, several different rates may be applied).

2.22   As a result of two decisions in the Supreme Court, it is no longer considered essential that the amount is calculated on the basis of a genuine pre-estimate of the loss likely to be suffered (*Cavendish Square Holdings* v *El Makdessi* and *ParkingEye Limited* v *Beavis*). Provided that the amount is not 'out of all proportion' to the likely losses, the damages will be recoverable without the need to prove the actual loss suffered, irrespective of whether the actual loss is significantly less or more than the recoverable sum (*BFI Group of Companies* v *DCB Integration Systems*). In other words, once the rate has been agreed, both parties are bound by it. Of course, for practical reasons, the rate should always be discussed with the client before inclusion in the tender documents, and an amount that will provide adequate compensation included, to cover among other things any additional professional fees that may be charged during this period. If 'nil' were to be inserted into the contract particulars then this would preclude the client from claiming any damages at all (see *Temloc* v *Errill*). If no sum were entered, the client may be able to claim general damages, but if this were the intention it would be better to set this out clearly rather than simply to leave the section blank.

2.23   If completion in sections is selected in item R, rates of liquidated damages are required for each section. Note that in this case, the rate should reflect the realistic losses to the client of not getting access to that section, which may not necessarily relate to the floor area of that part. In addition, the professional fees element may not be proportionally reduced.

Cavendish Square Holdings v *El Makdessi* and *ParkingEye Limited* v *Beavis*,
Supreme Court [2015] UKSA 67

In this landmark case the Supreme Court restated the law regarding whether a liquidated damages clause may be considered a penalty. Key criteria for whether a provision will be penal are: if 'the sum stipulated for is extravagant and unconscionable in amount in comparison with the greatest loss that could conceivably be proved to have followed from the breach'; and whether the sum imposes a detriment on the contract breaker which is 'out of all proportion to any legitimate interest of the innocent party'. In determining these, the court must consider the wider commercial context.

BFI Group of Companies Ltd v *DCB Integration Systems Ltd* [1987] ClLL 348

BFI employed DCB on the JCT Agreement for Minor Building Works to refurbish and alter offices and workshops at its transport depot. BFI was given possession of the building on the extended date for completion, but two of the six vehicle bays could not be used for another six weeks as the roller shutters had not yet been installed. Disputes arose which were taken to arbitration. The arbitrator found that the delay in completing the two bays did not cause BFI any loss of revenue, and that BFI was therefore not entitled to any of the liquidated damages. BFI was given leave to appeal to the High Court. HH Judge John Davies QC found that BFI was entitled to liquidated damages. It was quite irrelevant to consider whether in fact there was any loss. Liquidated damages do not run until possession is given to the employer but until practical completion is achieved, which may not be at the same time. Therefore, the fact that the employer had use of the building was also not relevant.

Temloc Ltd v *Errill Properties Ltd* (1987) 39 BLR 30 (CA)

Temloc entered into a contract with Errill Properties to construct a development near Plymouth. The contract was on JCT80 and was in the value of £840,000. '£ nil' was entered in the contract particulars against clause 24.2, liquidated and ascertained damages. Practical completion was certified around six weeks later than the revised date for completion. Temloc brought a claim against Errill Properties for non-payment of some certified amounts, and Errill counterclaimed for damages for late completion. It was held by the court that the effect of '£ nil' was not that the clause should be disregarded (because, for example, it indicated that it had not been possible to assess a rate in advance), but that it had been agreed that no damages would be payable in the event of late completion. Clause 24 is an exhaustive remedy and covers all losses normally attributable to a failure to complete on time. The defendant could not, therefore, fall back on the common law remedy of general damages for breach of contract.

Dispute resolution (item N, both contracts)

2.24   Three forms of dispute resolution are listed in the RIBA Building Contracts: adjudication, mediation and arbitration. In DBC, all three can be selected, or none, or any combination. In CBC, the adjudication option is preselected to ensure that the contract complies with the Housing Grants, Construction and Regeneration Act 1996 (as amended), which requires that all contracts covered by the Act include the right to adjudication. Otherwise, the same applies as for DBC, i.e. the parties may choose any combination.

2.25   It is important to give careful consideration to the choice of dispute resolution options. The differences between them are significant, and it might not be possible for one party to persuade the other to use a different system after a dispute has arisen; this therefore is the only opportunity to determine how matters will be resolved. In particular it should be noted that whereas adjudication is preselected in CBC, it is optional in DBC. This is because adjudication could be seen as favouring the contractor, given the short response times that may be difficult for a lay client to deal with. If selected, the provisions will nevertheless be binding (the courts have held that consumer clients are bound by an adjudication agreement in a standard form contract, as it does not significantly affect the balance of power between the parties) (*Lovell Projects* v *Legg & Carver*). The different options and their implications, including for the completion of item N, are discussed in Chapter 10.

> *Lovell Projects* v *Legg & Carver* [2003] BLR 452
>
> The employers were residential occupiers and consumers who engaged a contractor to refurbish their house using the JCT MW98. This edition included the JCT's own adjudication procedure (instead of relying on the Scheme for Construction Contracts, as it now does). The employers were advised by an architect, and during the tender negotiations they insisted that MW98 was used. Following an adjudication, the employers resisted enforcement of the decision on the basis that the adjudication provisions were unfair and had not been brought to their attention before they signed the contract. The court decided that contractual adjudication provisions in the consumer's contract were not unfair under the Unfair Terms in Consumer Contracts Regulations 1999 because they did not cause a significant imbalance in the parties' rights and obligations. The court bore in mind that it was the employers, not the contractor, who had proposed the term, and that the employers had access to advice from their contract administrator.

Contractor design (item P, both contracts)

2.26   If this item is selected it requires the client to indicate which aspects of the works are to be designed by the contractor. It is important that any aspect, whether a component, element or system (such as services), is delineated with care, as otherwise there may later be disagreements as to who is responsible for interfaces between these and any consultant-designed areas. It is very important that this entry is completed carefully and the consultant and client must give careful consideration to the issue of liability before sending out the tender documents.

2.27   As well as indicating which parts are to be designed by the contractor, the client is likely to want to stipulate requirements that the designed parts should meet. It would be sensible for the contract administrator to indicate in item P exactly where these requirements are set out (quite often this will be in the form of a performance specification). In addition, if the contractor is to be required to submit design proposals at tender stage, then this should be made clear in the tender documents, including the nature and extent of the submission required. Similarly, it may be sensible to set out details of any information to be submitted during the works, and the dates by which it will be required (cl. 15.2).

Insurance-backed guarantee and new building warranty (items U and V, both contracts)

2.28   These items require the client to set out details of any insurance-backed guarantee and/or new building warranty that will be required. These are similar products; the term 'new building warranty' is used mainly in the context of new homes, whereas 'insurance backed guarantee' can be applied to any type of building work. The National House Building Council's (NHBC) Buildmark scheme is a well-known example of a new building warranty, which covers homes built by NHBC-registered builders. The scheme protects the homeowner against defects for ten years; for the first two years the builder is responsible for putting right any defects, and during the third to tenth years any damage to a home resulting from building defects is covered by an insurance policy. It will also protect the homeowner if the builder goes bankrupt during the course of the work but before completion. Various trade associations offer a range of insurance-backed guarantees for domestic and commercial work, for example the Federation of Master Builders' Build Assure scheme, which offers a variety of packages, from simple liquidation cover to ten years' protection against structural damage. If selected, the insurance-backed guarantee is to be provided before the start date (cl. 20.1).

Collateral warranty/third-party rights (item X, CBC only)

2.29   A collateral warranty is a separate contract entered into by the contractor and a third party, typically a purchaser or tenant of the completed building or a project funder. They are particularly important if the contractor is undertaking design, as without one the purchaser, tenant or funder will not be able to recover its losses should the building later develop defects.

2.30   If the client wishes the contractor to provide such a warranty, this should be made clear in the tender documents (if not the contractor may later agree to provide one, but the client would not be able to insist that it does). The details of to whom the warranty is to be provided and the form of warranty to be used should also be given. Standard forms of warranty are published by the JCT, although these may require some modification for use with the RIBA Building Contracts. The contractor must execute the warranties as set out within 14 days of being requested to do so by the client (cl. 23.1).

2.31   As an alternative to the use of a collateral warranty, the contract refers to the use of a third-party rights agreement. This facility was introduced by the Contracts (Rights of Third Parties) Act 1999. Until this Act came into force, it was a rule of English law that only the two parties to a contract had the right to bring an action to enforce its terms (termed 'privity of contract'). Now, the Act entitles third parties to enforce a right under a contract, where the term in question was intended to provide a benefit to that third party. The third party could be specifically named or it could be an identified class of people. The Act, however, allows for parties to agree that their contract will not be subject to its provisions, and many standard forms adopt this course in order to limit the parties' liabilities. Both RIBA Building Contracts state: 'Third parties have no rights under the Contract' (cl. 11.4). The concise contract, however, states that this is 'unless item X of the Contract Details has been selected' and provides for the client to require the contractor to enter into a 'Third Party Rights Agreement' (cl. 23.1), details of which must be set out in the Contract Details. It should be noted that there are no standard forms of agreement available for use with the RIBA Building Contracts, therefore the parties would need to draft their own.

Pre-contract negotiations

2.32   Once the tenders have been returned, the next stage is to finalise the agreement with the selected contractor. Ensuring that a contract is in place before construction work begins is a key task of the contract administrator, and to neglect to do so may constitute negligence.

2.33   A contract is formed when an unconditional offer is unconditionally accepted. In the context of a building project, where contractors have been invited to submit competitive tenders, the tenders constitute 'offers' to carry out the work shown in the tender documents for the price tendered. If a tender is accepted unconditionally then a contract will have been formed, and the terms of the contract will be those set out or referred to in the tender documents.

2.34   'Letters of intent' can cloud the picture and should be avoided. If it is possible to accept the tender without qualification then it is better simply to write a letter to that effect, and the contract comes into existence from the moment the letter has been received by the contractor.[2] The effect of a letter expressing an intention to enter into a contract at some point in the future will depend on the wording and circumstances in each case, but it is likely to be of no legal effect. Starting work on such a basis could have disastrous consequences for both parties.

2.35   If there is a period of negotiation before the formal contract is drawn up, careful records should be kept of all matters agreed in order that they can be accurately incorporated into the formal contract documents. These documents should always be prepared as soon as an agreement is reached, and before work commences on site. Failure to execute the documents does not necessarily mean that no contract is in existence, but it might give rise to sufficient doubt to require spending on legal fees in an effort to establish the true position (*Goldsworthy* v *Harrison*). In addition, it can often lead to avoidable arguments about what was agreed.

2.36   The formal contract, once executed, will supersede any conflicting provisions in the accepted tender and will apply retrospectively (*Tameside Metropolitan BC* v *Barlow Securities*).

Goldsworthy and *others* v *Harrison* and *another* [2016] EWHC 1589 (TCC)

The case concerned the enforcement of an adjudicator's decision. The defendant employers were residential occupiers and the key matter in dispute was whether the parties had agreed contract terms that contained an adjudication clause. If they had not, the adjudicator had no jurisdiction. At the time of tender, there had been no mention of the use of JCT MW. However, an email around the time work began stated: 'As discussed previously the contract will be a JCT Minor Work', which was followed shortly after by another stating: 'If you are successful in the quotation for the garage and summer house, and main works please note at that time there will be retentions applied and the JCT Minor Works Contract'. However the documents were never executed, and there was no evidence that key matters such as liquidated damages had been agreed. The court refused to give summary judgment, stating that without fuller evidence from both sides, in particular of the discussions lying behind the emails, it was impossible to say that there was not a triable issue on the question of whether the parties had agreed on the JCT Minor Works terms, in particular the gaps where particular options had not been filled in or agreed. The matter would therefore have to proceed to a full trial.

---

[2]   For a more detailed analysis of the formation of contracts the reader could refer to Stephen Furst and Vivien Ramsey (eds), *Keating on Construction Contracts*, 9th edn (London: Sweet & Maxwell, 2012).

Tameside Metropolitan Borough Council v *Barlow Securities Group Services Limited* [2001] BLR 113

Under JCT63 Local Authorities, Barlow Securities was contracted to build 106 houses for Tameside. A revised tender was submitted in September 1982 and work started in October 1982. By the time the contract was executed, 80 per cent of building work had been completed, and two certificates of practical completion were issued relating to seven of the houses in December 1983 and January 1994. Practical completion of the last houses was certified in October 1984. The retention was released under an interim certificate in October 1987. Barlow Securities did not submit any final account, although at a meeting in 1988 the final account was discussed. Defects appeared in 1995, and Tameside issued a writ on 9 February 1996. It was agreed between the parties that a binding agreement had been reached before work had started, and the only difference between the agreement and the executed contract was that the contract was under seal. It was found that there was no clear and unequivocal representation by Tameside that it would not rely on its rights in respect of defects. Time began to run in respect of the defects from the dates of practical completion; the first seven houses were therefore time barred. Tameside was not prevented from bringing the claim by failure to issue a final certificate.

Preparing and executing the contract package

2.37   The RIBA Building Contracts have an 'Agreement' section at the start of the document that is to be signed by both parties. The contracts can be executed as a simple contract (sometimes termed 'under hand' in other contracts) or as a deed; if neither option is selected, the default is a simple contract. It is good practice to sign not only the form itself, but also a copy of all other contract documents, to avoid any arguments about exactly which contract or revisions are included in the contract. In CBC the client is given the option of signing as an individual or as a registered company; obviously, if it signs as an individual, the client will be personally liable to the contractor, whereas as a company its liability will be limited (see *Hamid* v *Francis Bradshaw Partnership*). The option of signing as a company is not offered in DBC, because a company would not fall under the residential occupier exception discussed at para. A1.7 of appendix 1. Therefore, DBC cannot be used by, for example, a homeowner who wishes to engage a contractor through a company they own.

2.38   Once the tender process is complete and the contract has been signed, the contract administrator is required to give the contractor 'up to two copies of the Contract Documents' (cl. 5.2.1). This could be in any format the parties agree; there is no need to sign multiple sets – one set should be signed and then copies made. It may be that the contractor would prefer to have one of these as a secure PDF. The original is usually retained by the client, with the contract administrator holding a copy (or the other way around). Any changes are to be issued to the parties (cl. 5.2.2); this might happen, for example, if the parties agree an amendment. Any amendment should be in writing and signed by both parties (cl. 11.7).

Interpreting the contract and resolving inconsistencies

2.39   The RIBA Building Contracts are clearly laid out and generally easy to understand. However, it should be remembered that the contract between the parties comprises not only the signed contract form, but all the contract documents as well. Even if great care has been taken in preparing the contract documents, there can be unintended conflicts or inconsistencies between them. As a result, it may be difficult to anticipate their combined effect. This section looks at how the contract itself and the package of contract documents are to be interpreted, and how any errors or problems are corrected.

Definitions

2.40   A 'Definition of Terms' is set out on pages 17 and 18 of both versions of the contract. These should be used to interpret the meanings of clauses (they would be used, for example, by a court). For example, a Payment Notice is defined as 'a notice that the Contractor issues to the Client, in accordance with clause 7, showing the payment that the Contractor considers is due and how it was calculated'. The contract also includes 'defined terms' that are not included in the Definition of Terms but are defined within the Contract Details. These terms are all capitalised throughout the contract and include, for example, 'Contractor', 'Client' and 'Start Date'.

2.41   Note that no definition of 'a day' is given. It is suggested that in calculating periods of days, parties should adopt the normal procedure that all days including weekends should be counted, except for public holidays. To avoid any passible disputes, it may be sensible to set this definition in the specification preliminaries.

Priority of contract documents

2.42   Clause 11.1 states that:

> All parts of the Contract shall be read together as a whole; however in all circumstances the Agreement, Contract Details and these Contract Conditions shall take precedence over all other Contract Documents.

This means that conflicting provisions in other documents, for example in the preamble to a bill of quantities, will not override the printed conditions. This gives clarity and avoids unintended clashes occurring. However, if the parties wish to agree special terms that differ in any way from the printed conditions, then it will not be sufficient to simply append them in a separate document: amendments must be made to the actual printed contract (note: this is easier to do in the online version). This could be done by amending the individual clauses in the printed contract; alternatively, clause 11.1 could be amended by adding a qualification that the revisions set out in another document will take precedence over the printed contract. Amending standard contracts is unwise without expert advice, as the consequential effects are difficult to predict. Deleting clause 11.1 would be particularly unwise as it may have unintended effects on the contract as a whole.

2.43   Subject to clause 11.1 (i.e. that the printed conditions take precedence), the parties would be able to set out a priority between the other contract documents. If no priority is set out and a conflict exists that gives rise to a dispute, the adjudicator or court would need to decide, on a balance of probabilities, what the parties' intentions were when they made the contract. This is done objectively, i.e. the adjudicator or court is not interested in the subjective intention, but in what a disinterested bystander would conclude the parties had meant. A contract administrator trying to resolve conflicts in this situation would need to be equally objective, i.e. when looking at the documents as a whole, what, on balance, do they appear to mean?

Inconsistencies, errors or omissions

2.44   Clause 5.9 states:

> If the Contractor finds any inconsistency in the Contract Documents and/or an instruction, it shall inform the Architect/Contract Administrator immediately. The Architect/Contract Administrator is to issue instructions to resolve the inconsistency.

The equivalent clause in the 2014 editions also required the contractor to notify the contract administrator of inconsistencies 'between the documents and the law'. It is not clear why this was removed, but it would normally be implied that a competent contractor ought to notify any such discrepancies it finds, especially if the shortfall is a serious matter.

2.45   The contract administrator is given the power under clause 5.4.3 to issue instructions to correct any inconsistency in the contract documents, which would cover issues that have not been notified. Although expressed as a power, and not a duty (as in JCT forms), it is suggested that the contract administrator should normally take steps to resolve any problems that emerge. Any such instruction that results in a change to works is to be dealt with in the same way as a clause 5.4.1 change to works instruction (cl. 5.11, see paragraph 6.12).

3 Roles and management systems

3.1 Contracts perform many functions: they assign risk, they set out obligations and rights, and they can also act as a management tool.

3.2 Generally speaking, where a contract states that an action 'shall' be performed, this indicates an obligation, also referred to as a 'duty'; where it says it 'may' be performed, this is an optional action, referred to as a 'right' or a 'power'. Some of the obligations and rights could be described as core; for example, the contractor's duty to complete the works, the client's duty to pay the contractor, and the contract administrator's right to instruct a change to the works. Some are procedural, for example the obligation to submit a document within a particular time frame. The RIBA Building Contracts additionally have a group of clauses, under the heading 'Collaborative Working', which are aimed at assisting the smooth management of the contract.

3.3 This chapter describes the roles of the contract administrator, the contractor and the client, outlining their duties and rights. It then looks at some of the provisions in the RIBA Building Contracts that aim to ensure best practice management procedures are applied to projects.

Role of the contract administrator

3.4 The contract administrator is appointed by the client, and its duties and liabilities are owed to the client as set out in its terms of appointment. The contract administrator is not a party to the contract but is named in the contract (cl. 5.1), and the extent of its authority to administer the contract derives from the wording of the contract.

3.5 The RIBA Building Contracts place various duties on the contract administrator, particularly with regard to issuing certificates or statements, as well as a wide variety of powers, such as the power to issue instructions. For a full list of these duties and powers, along with references as to where they are discussed in this Guide, see Tables 3.1 and 3.2.

3.6 In some matters the contract administrator will act as an agent of the client, for example when issuing instructions that will vary the works. In other instances it acts as an independent decision-maker, e.g. when deciding on claims for additional payment. When acting in the latter capacity, it would be implied that the contract administrator must act fairly at all times. It would be sensible for the contract administrator to use one of the RIBA standard forms of appointment to ensure that the obligations under its appointment align with those in the RIBA Building Contracts.

3.7 Failure by the contract administrator to comply with any obligation, either express or implied, may result in the contractor suffering losses. As the contract administrator is not a party to the contract, if the contractor wishes to bring a claim, this would, in the first instance, have to be against the client. It is likely, however, that any failure to administer the building contract according to its terms would be a breach of the contract administrator's duties to the client and, therefore, the client may seek, in turn, to recover its losses from the contract administrator.

Table 3.1 Contract administrator's duties

Clause		Duty
CBC	DBC	
3.1	3.1	Attend the pre-start meeting
3.3	3.3	Take into account (when determining a revision of time and/or additional payment) any failure by the contractor to comply with 3.2
5.1	5.1	Administer the contract, issuing instructions and certificates and taking decisions
5.2.1	5.2.1	Give the contractor up to two free copies of the contract documents
5.2.2	5.2.2	Issue any revisions
5.7	5.7	Confirm oral instructions to the contractor in writing
5.9	5.9	Issue instructions to deal with inconsistencies in contract documents
5.12	5.12	Aim to agree any revision of time and/or additional payment promptly
5.13	5.13	Determine the appropriate adjustment to the contract time and/or price
7.3	7.3	Issue payment certificates, not later than 5 days after the date for interim payment
7.11.2	7.11.2	Aim to agree the final contract price with the contractor within 30 days of the contractor's submission
7.11.3	7.11.3	Calculate the final contract price, if the parties cannot agree or the contractor makes no submission, and notify the parties in writing
7.12.1	7.12.1	Issue a final payment certificate
9.5	9.5	Aim to agree on a revision of time promptly
9.6	9.6	Make a reasonable assessment of a revision of time, if unable to agree
9.8	9.8	Aim to agree on an additional payment promptly
9.9	9.9	Make a reasonable assessment of any additional payment, if unable to agree
10.4	10.4	Issue a certificate of making good defects when satisfied that defects are fixed
10.5	10.5	Issue the contractor with a notice if it fails to fix a defect
18.3	18.3	Notify the parties when satisfied that a milestone has been achieved

Table 3.2	Contract administrator's powers	
Clause		**Power**
CBC	**DBC**	
5.3	5.3	Visit the site, including any off-site locations in connection with the works
5.4.1	5.4.1	Instruct a change to the works
5.4.2	5.4.2	Postpone the works or a section
5.4.3	5.4.3	Resolve inconsistencies in the documents or an instruction
5.4.4	5.4.4	Reject defective work
5.4.5	5.4.5	Require further documents
5.5	5.5	Instruct work is uncovered and inspected or tested
5.8	5.8	Issue a 7-day notice to comply with an instruction
5.15	5.15	Modify, amend or withdraw an instruction, following a cl. 5.14 notice from the contractor
9.11	9.11	Certify practical completion
12.1	12.1	Issue the contractor with a notice of intention to terminate
12.2	12.2	Issue the contractor with a notice of termination

Role of the client

3.8 The client has a significant role in any construction project. Its key duties, outlined in Table 3.3, are ones of collaboration. For example, not to take any action that would interfere with the carrying out of the works, and to pay the contractor on time in accordance with the contract. The client is also given various rights, as listed in Table 3.4. Generally, however, the client should discuss both its duties and its rights with the contract administrator, and should seek the contract administrator's advice before taking any action; the contract administrator has the overall responsibility for administering the project, and if actions are coordinated there is less likelihood of clashes or unexpected consequences.

Table 3.3	Client's duties	
Clause		**Duty**
CBC	**DBC**	
1.1	1.1	Allow contractor access to the site for pre-construction inspection, carrying out the works and rectifying defects
1.2	1.2	Provide the contractor with access to parts of the works taken over prior to practical completion
1.3	1.3	Inform the contractor of any other appointments
1.5	1.5	Comply with all relevant health and safety legislation
3.1	3.1	Attend the pre-start meeting

Table 3.3 Continued

Clause		Duty
CBC	DBC	
3.2	3.2	Provide the contractor and the contract administrator with a warning of an event that will affect progress of the works, and work together with the contractor to resolve the event. Take reasonable steps to minimise the effect of the event
6.3	6.3	Maintain insurance in respect of its liabilities to the value specified
6.4	6.4	Provide evidence of insurance, take out insurance if the contractor fails to provide evidence
7.6	7.6	Pay amount due on any payment certificate or payment notice by the final date for payment
7.15.2	7.15.2	Pay the contractor VAT on presentation of a valid VAT invoice
9.1	9.1	Inform the contractor if force majeure occurs
19.1	–	Pay the contractor an advance payment
22.1	–	Provide evidence of its ability to pay the contract price

Table 3.4 Client's rights

Clause		Right
CBC	DBC	
1.4	1.4	Defer access to the site or sections of the site
5.8	5.8	Employ others if the contractor fails to comply with a 7-day notice
7.9	7.9	Issue a pay less notice
9.12	9.12	Request to take over part of the works before practical completion
10.1	10.1	Deduct liquidated damages
10.6	10.6	Employ others if the contractor fails to rectify a notified defect
12.6	12.6	Terminate the contractor's employment with a termination notice if the works are suspended by 60 days
–	12.11	Cancel the contract within 14 days
13.1	13.1	If mediation is selected, refer disputes to mediation
13.2	13.2	If adjudication is selected, refer disputes to adjudication
13.4	13.4	If arbitration is selected, refer disputes to arbitration
13.5	13.5	If arbitration is not selected, refer disputes to the appropriate court

Client acting as contract administrator

3.9 DBC is unusual in that it offers the facility for the client to name themselves as contract administrator (optional cl. 19). Although it is sometimes frowned upon, there is no reason why the client can't name themselves in this way, with the very important proviso that it must have been made completely clear to the contractor at tender stage that this was going to be the arrangement. The client must also ask themselves whether they have the skills and resources (particularly time) necessary to perform the role, and whether they are capable of being truly impartial. Any incorrect or biased decisions are liable to be challenged by the contractor and, if a reasonable solution is not agreed, may result in the contractor being successful in subsequent litigation (as the tribunal is likely to examine the decision very closely). It is unwise for a client to hope that it may secure a 'better deal' by taking on the role of contract administrator.

Role of the contractor

3.10 The contractor has the most extensive lists of duties and rights, which is unsurprising as it is the contractor who is primarily responsible for delivering the project on time and to the client's requirements. Full lists of these duties and rights are provided in Tables 3.5 and 3.6, and the key duties are outlined below.

Completing the works

3.11 The main obligation of the contractor is to complete the works as set out in the contract documents. In both CBC and DBC, this obligation is set out in clause 2.1, which states:

> The Contractor shall:
>
> 2.1.1 carry out and complete the Works in accordance with the Contract, in good and workmanlike manner, by the Date for Completion
>
> 2.1.2 be responsible for all regulatory and statutory consents, fees and charges as set out under item I of the Contract Details
>
> 2.1.3 comply with all of the relevant health and safety legislation
>
> 2.1.4 comply with all statutory obligations applicable to the Works.

3.12 It is essential that the standards for materials, goods and workmanship required for the works are set out clearly in the specification and other contract documents. The above clause does not refer expressly to 'materials, goods and workmanship', as would be usual in JCT contracts, nor does the contract set any further requirements for these. It is suggested that 'in good and workmanlike manner' refers to the method of carrying out the work, not what work is to be done. Clause 2.2 adds a further requirement as to working methods, stating that 'the Contractor shall use methods that prevent nuisance, trespass and pollution'.

Contractor's design obligation

3.13   Like most traditional contracts these days, the RIBA Building Contracts make provision for some design to be carried out by the contractor. It is an optional provision, but it is likely that on most projects it will be required to some extent. Any decision that affects the final form of the building (as opposed to the method of construction) is a design decision, even if it relates only to small levels of detail, such as the exact size of joist hangers or of central heating pipes. If in any doubt as to whether something is effectively 'design', the contract administrator should consider whether it is intending to make all the necessary decisions, and, if not, what would happen if the particular detail were to fail: would it be prepared to accept responsibility? If not, it should take steps to ensure that liability for design of that detail is expressly placed with the contractor.

3.14   The provisions are set out in optional clause 15, and the parts to be designed by the contractor are described in item P of the Contract Details (both contracts). A 'Contractor's design proposal' is referred to in item D, therefore the contract anticipates that the contractor may have provided a design prior to the contract being executed, which may be for some or all of the parts, depending on what was requested at tender stage (see paras 2.15, 2.16 and 2.27). It appears that the contractor is responsible for designing only the parts that are listed. Although there is no clause stating that the contractor is not liable for design provided by the client or the client's agents, as there was in the 2014 edition (A2.4), it is unlikely that a court would find a contractor liable for the contract administrator's design. However, the contractor must notify the contract administrator of any discrepancy it finds between its designs and those of the contract administrator (cl. 15.3).

3.15   What happens if the contractor makes a design decision relating to a part that has not been stipulated in the Contract Details, but for which the contract administrator has provided no design information? It is possible that the contractor would not be held responsible for that design decision. A relevant case concerns the Museum of Liverpool (*National Museums and Galleries on Merseyside v AEW Architects and Designers*), where the judge said the contractor was not responsible for designing anything not identified as part of the contractor's designed portion (CDP), in this case deciding the tolerance gaps between steps. Courts might a take less strict view in a smaller project, but to be safe it is best to use the 'Contractor Design' optional clause and to make sure the extent of the contractor's design responsibility is clear.

National Museums and Galleries on Merseyside v AEW Architects and Designers Ltd [2013] EWHC 2403 (TCC)

This project, let on SBC05, was for a new museum, constructed between 2007 and 2011. A key design element was a series of 'half amphitheatre' pre-cast concrete steps and seats at the north and south ends of the museum.

Unfortunately, architects AEW made several errors in coordinating the detailed design of the project, including the valley junction between the concrete steps and seats. The steel substructure to these had been redesigned by the engineers (Buro Happold) in August 2007, and AEW failed to appreciate the implications this had for the geometry of the interface between the steps and seats, or to specify the dimensional tolerances between the pre-cast units, or an adequate coverage for the reinforcement to the units, even after they were alerted to the problems in 2008 by a query from the contractor. As a result, it was not possible to use the steps at the time the museum was opened to the public in 2011. The problems resulted in a claim by the museum against the architects, who tried to argue that this detail was the contractor's responsibility.

However, the judge would not accept this argument, stating (at para. 82):

> In relation to the gaps, AEW suggests that the design of the steps and seats was part of the works which the Contractor was required to design. This is, simply, wrong. The construction contract identifies those parts of the Works which the Contractor was required to design or have design involvement with as: 'steelwork connections, reinforcement placement & scheduling, general glazing & curtain walling, roof cladding, fixing wind posts, structural glass and glazing'. This is described in the contract as the 'Contractor's Designed Portion' and it is simply in relation to those works that the Contractor has any design responsibility.

3.16   It is therefore very important that the delineation between the parts to be designed by the contractor and the rest of the project is described accurately.[1] This can be quite difficult in practice, especially where several parts or elements are listed, or the list includes a system (e.g. services) that is integral to many parts of the building.

3.17   As the contractor is only required to design the specified parts, the contract administrator will remain responsible for any integration. This could extend to the physical junctions between the contractor-designed parts and other parts, but could also cover the combined performance of several systems or of systems and elements. If it is intended that the contractor is to be responsible for resolving any interface (physical or performance), the interface would have to be identified clearly within the parts to be designed by the contractor. The contractor retains the copyright in any design it provides, but grants the client a licence to use it for the Works and related purposes (cl. 15.4).

Contractor's design liability

3.18   Under clause 15 the level of liability of the contractor for design is to use the skill and care of a competent designer (cl. 15.1.1). The contractor is also required 'to ensure that its design is in accordance with the Client's specification as stated in the Contract Document' (cl. 15.1.2).

3.19   There are essentially two levels of liability used in construction. Professionals normally undertake to use reasonable skill and care, rather than promise to achieve a particular result. To understand this distinction, consider the case of a medical professional: a doctor would never promise to cure a patient, but simply to use their medical skills competently to achieve the best outcome possible. The fact that the patient does not get better is not of itself sufficient to show the doctor made a mistake. Similarly, with professional designers such as architects, the fact that there is a defect in the design is not sufficient to show they were negligent. Any client bringing a claim must also prove that the architect failed to use the reasonable skill and care of a competent architect. This is sometimes referred to as a 'negligence-based liability' and is the normal level of liability that would be set out in an architect's appointment or would be implied by the courts if no level has been set.

---

[1]   An example of the confusion that can arise when the contractor-designed items are not accurately defined or described can be seen in *Walter Lilly & Company Limited* v *Giles MacKay and DMW Ltd*.

3.20   The alternative, more onerous, level of liability is to promise to achieve a result. This is often referred to as a 'strict liability', as there is no need for the claimant to show that there was any negligence on the part of the designer; it would be enough to show that the result did not meet the stated requirements in some respect. A commonly used shorthand descriptor is that the designer has taken on a 'fit for purpose' liability. Contractors who undertake design work often take on an express 'fit for purpose' liability. Where no level of liability had been set out, the courts may imply that there is a strict liability, provided it can be shown the client was relying on the contractor's skill and judgement.

3.21   Clause 15 appears to contain both levels of liability, i.e. a professional one under 15.1.1 and a strict one under 15.1.2 – the phrase 'ensure that its design is in accordance with…' would normally be interpreted as a strict duty. This could potentially create a conflict, however, the 15.1.2 obligation is stated to be 'subject to clause 15.1.1' and so it is likely that the former level would prevail. Nevertheless, this may depend on how the rest of the documents are expressed (see, for example, the recent case of *MT Højgaard A/S v E.ON Climate & Renewables UK Robin Rigg East Ltd*).

MT Højgaard A/S v E.ON Climate & Renewables UK Robin Rigg East Ltd

Here MT Højgaard (MTH), the claimant contractor, entered into an agreement with the defendant employers E.ON Climate and Renewables (EON) for the design, fabrication and installation of the foundations for 60 wind turbine generators for the Robin Rigg offshore wind farm in the Solway Firth, Scotland. Following installation of the turbines, movement was discovered in the grouted connections between the foundation monopoles and the transition pieces which supported the generators. The grouted connections had been designed in accordance with international standard DNV-OS-J101, as required by the specification. The key clauses stated:

8.1 GENERAL OBLIGATIONS
The Contractor shall, in accordance with this Agreement, design, manufacture, test, deliver and install and complete the Works:

(i)   with due care and diligence expected of appropriately qualified and experienced designers, engineers and constructors (as the case may be).

…

(viii)   so that the Works, when completed, comply with the requirements of the Agreement.

…

3.2.2.2 Detailed Design Stage
The detailed design of the foundation structures shall be according to the method of design by direct simulation of the combined load effect of simultaneous load processes (ref: DNV-OS-J101). Such a method is referred to throughout this document as an 'integrated analysis'

The design of the foundations shall ensure a lifetime of 20 years in every aspect without planned replacement. …

Unfortunately, DNV-OS-J101 was fundamentally flawed, so that following it would have inevitably resulted in a faulty connection. The court therefore had to decide whether MTH's obligation was limited to using reasonable skill and care (i.e. to design the foundations on the basis of a 20 year design life in accordance with J101) or whether it was under a strict obligation to achieve a service life of 20 years. The Technology and Construction Court (TCC) decided that the words of clause 3.2.2.2 were clear and imposed a strict obligation, noting that it is not uncommon for the obligations to exercise reasonable care and to achieve a particular result to exist side by side in construction and engineering contracts, and that the two obligations are not mutually incompatible. Although this was overturned at the Court of Appeal, it ultimately proceeded to the Supreme Court, where the original decision was upheld.

3.22   Clients would usually prefer a strict level of liability. After all, if they have taken the trouble, with their consultant, to set out detailed requirements for the design, including specific performance targets (e.g. in relation to energy use), they are likely to want to be able to bring a claim should the building not perform, without having to also prove a lack of care. However, as the 'fit for purpose' level of liability is more onerous on the contractor, some contractors may be unwilling to tender on this basis or might submit higher tender prices. The JCT contracts do not generally contain this option (except in the case of the Major Project Construction and Constructing Excellence contracts). However, it is the default position in NEC4 contracts, with the alternative of 'reasonable skill and care' under Option X15, and is standard in the FIDIC and IChemE contracts, therefore its use is increasing.

3.23   Under clause 15.5 the contractor is required to 'ensure there is adequate professional indemnity insurance for its design responsibilities, as set out in item P of the Contract Details' (if no requirement is set out the obligation would fall away). The contractor might not be able to obtain insurance to cover a 'fit for purpose' risk, but that would not, of course, affect its liability.

Contractor's obligations in respect of subcontracted work

3.24   The contractor's duties with respect to subcontracting the work are set out briefly in the following clause:

> 2.6   The Contractor shall be solely responsible for carrying out the Works and for the performance of all subcontractors and suppliers, including any Required Specialists listed in item Q of the Contract Details (if selected).

3.25   The contractor may subcontract to anyone it chooses; there is no requirement to inform the contract administrator beforehand, or to obtain permission from the contract administrator or the client, and neither has the power to bar a contractor from using a particular firm. As the contractor is, of course, entirely responsible for all its subcontractors, in theory there is no risk to the client, even if the contractor selects firms that are not capable of achieving the desired standard of work. However, in practice it can be frustrating to stand by without intervening, as work has inevitably to be redone and delays are caused.

3.26   If the client wishes to take a more proactive stance, there are two possible courses: one is to add a clause requiring the client's or contract administrator's permission for any subcontracting (or perhaps only for subcontracting certain work), with such permission not to be unreasonably withheld. The second is to use the 'Required Specialists' provisions for critical areas of work.

Required specialists

3.27   Under an optional clause (cl. 16) the client may specify that the contractor uses particular specialists to carry out described parts of the works in the contracts. If used, the contractor is fully responsible for the performance of those specialists (cl. 16.1.1), in the same way that it would be for its own domestic subcontractors.

3.28   The details of the specialists, and the work they are to carry out, are entered in the Contract Details (item Q). The contractor should be given full information at tender stage to allow the contractor to make enquiries as to the suitability of the firm, and its price and terms for carrying out the work.

3.29   Clause 16.2 states: 'The Contractor shall, at the Start Date or thereafter as appropriate, employ the Required Specialists to undertake the part(s) of the Works described in item Q of the Contract Details.'

3.30   It is possible that if the contractor experiences problems with the specialist, it may later refuse to engage the firm, or engage the specialist, but then attempt to claim for additional time and money should the firm later default on its subcontract (there is old case law that may support such a claim, for example see *Gloucestershire County Council v Richardson*). It is suggested that in the context of the RIBA Building Contracts any such argument is unlikely to succeed: clauses 2.6, 16.1.1 and 16.2 are unqualified – it is the responsibly of the contractor to make full enquiries and request any further information at tender; once the contract is entered into, the specialist's performance is entirely at the contractor's risk.

> *Gloucestershire County Council v Richardson* [1969] 1 AC 480 HL
>
> In this case the House of Lords considered whether a main contractor might be liable to the client in respect of latent defects in materials delivered by a nominated supplier. It held that the main contractor's liability to the client was limited to the extent of the nominated supplier's liability to the main contractor by operation of the terms of the nominated subcontract.

3.31   If the contractor terminates the required specialist's employment, it must notify the contract administrator, and provide a list of suitable alternative specialists (cl. 16.1.2). The contract administrator is required to 'issue an instruction regarding a replacement' (cl. 16.1.3). The contract then states that the contractor 'shall be responsible for any delay or additional costs arising from the termination' (cl. 16.3). Essentially, the contract administrator is responsible for taking action and resolving the situation. However, as the contractor will bear all the losses, it is likely to make every effort to make sensible suggestions for a replacement.

3.32   The contracts do not distinguish between the possible different reasons for the termination, so it appears that the liability would apply whatever the reason; whether a simple falling out or a disagreement about money, or because the firm seriously failed to perform. Furthermore, they do not discuss what would happen if the required specialist refused to undertake the project, or became insolvent, and yet no termination was issued. The contractor is only made liable for consequences of any termination, not of complete non-performance, and a contractor might argue that in these circumstances the client should carry the risk. It would therefore be sensible for the contract administrator to take action promptly, whatever the cause of the termination.

Compliance with statute/health and safety legislation

3.33   The contractor is required to comply with all statutory requirements applicable to the works (cl. 2.1.4). In addition, clause 2.1.3 requires it to 'comply with all of the relevant health and safety legislation'. The contracts also require the contractor to notify the contract administrator if, in its opinion, an instruction may have adverse health and safety implications (cl. 5.14.2). This is a useful provision that reflects the normal implied duty to warn of any instruction that would have an adverse effect.

3.34   Clause 2.1.2 states the contractor shall: 'be responsible for obtaining all regulatory and statutory consents, fees and charges as set out under item I of the Contract Details'. It is important to note that it is not just a matter of fees, notices and charges as would be the case in many JCT contracts; item I could list responsibility for obtaining and paying for all planning permissions, Building Regulations approvals and/or party wall consents.

3.35   Such matters would, of course, place far greater responsibility than the general compliance set out in clause 2.1.4. If the client wishes the contractor to obtain such consents, it would be important to make this clear in the tender documents, and to ensure that 'Contractor' is selected in item I, as otherwise this would not be covered by the lesser obligation in clause 2.1.4.

Table 3.5 Contractor's duties		
Clause		**Duty**
CBC	**DBC**	
2.1.1	2.1.1	Carry out and complete the works in accordance with the contract, in good and workmanlike manner, by the date for completion
2.1.2	2.1.2	Be responsible for all regulatory and statutory consents, fees and charges as set out under item I of the Contract Details
2.1.3	2.1.3	Comply with all relevant health and safety legislation
2.1.4	2.1.4	Comply with all statutory obligations
2.2	2.2	Use methods that prevent nuisance, trespass and pollution
2.3	2.3	Ensure a suitably qualified representative is available during the works
2.4	2.4	Take all reasonable steps and precautions to ensure that security is maintained on the site at all times
3.1	3.1	Attend a pre-start meeting
3.2	3.2	Provide the client and the contract administrator with a warning of any event that will affect progress of the works, and work together with the client to resolve the event, and take reasonable steps to minimise its effects
4	4	Obtain written consent of the client before assigning rights or benefits
5.6	5.6	Comply with instructions promptly
5.8.2	5.8.2	Cooperate with others engaged by the client following contractor failure to comply with a 7-day notice
5.8.3	5.8.3	Allow others as above access to the site
5.8.4	5.8.4	Be responsible for all costs incurred

Table 3.5 Continued

Clause		Duty
CBC	DBC	
5.9	5.9	Inform the contract administrator immediately of any inconsistencies it finds in the contract documents and/or any instruction
5.11	5.11	Calculate the effect of a change to works instruction on contract price and completion date
5.12	5.12	Aim to agree any revision to time or contract price promptly
5.14	5.14	Inform the contract administrator within 7 days if it believes that any instruction is not in accordance with the contract, or would have adverse effects on health and safety or the contractor's design
5.15	5.15	Comply with the contract administrator's decision regarding the above
6.3	6.3	Maintain insurance as set out in the Contract Details
6.4	6.4	Provide evidence of insurance, or take out insurance if the client fails to provide evidence
7.6	7.6	Pay the client any balance shown as due on the interim payment certificate
7.11.1	7.11.1	Submit its calculation of the final contract price, along with supporting documentation, not later than 90 days after practical completion
7.11.2	7.11.2	Aim to agree the final contract price with the contract administrator within 30 days of its submission
7.15.1	7.15.1	Issue the client with a valid VAT invoice
9.1	9.1	Inform the client if force majeure occurs
9.4	9.4	Inform the contract administrator of any event affecting the progress of the works and apply for a revision of time (with supporting documentation) within a reasonable time
9.5	9.5	Aim to agree on a revision of time promptly
9.7	9.7	Apply for any adjustment to the contract price within a reasonable time of an event occurring
9.8	9.8	Aim to agree on an additional payment promptly
9.11	9.11	Notify the contract administrator when it thinks that practical completion of the works or a section has been achieved
10.3	10.3	Remedy all defects identified during defects fixing period
12.10.2	12.10.2	Remove all materials and equipment from site
14.1	14.1	Submit a programme to the contract administrator, in the agreed format, no later than the pre-start meeting
15.1.1	15.1.1	Use reasonable skill, care and diligence in the design
15.1.2	15.1.2	Ensure the design is in accordance with the client's specification
15.2	15.2	Submit details of its design to the contract administrator
15.3	15.3	Notify the contract administrator of any discrepancies it finds in the client or architect's designs
15.5	15.5	Ensure there is adequate professional indemnity insurance for its design

Table 3.5 Continued

Clause		Duty
CBC	DBC	
16.1.2	16.1.2	Notify the contract administrator of the termination of the employment of a required specialist
16.2	16.2	Employ the required specialists to carry out the described parts of the works
20.1	20.1	Provide evidence of an insurance-backed guarantee
21.1	21.1	Provide a new building warranty
23.1	23.1	Execute a collateral warranty or third-party rights agreement in favour of identified parties
24.1	–	Adhere to all legislation relating to official secrets
24.2	–	Pass all requests for information under the Freedom of Information Act 2000 to the client
24.4.1	–	Ensure that discrimination in any form is not practised or allowed

Table 3.6 Contractor's rights

Clause		Right
CBC	DBC	
2.5	2.5	Use, free of charge, any client facilities listed in the contract documents
3.4	3.4	Propose changes to the works that may improve their value or lower the contract price
7.2	7.2	Send the contract administrator an application for payment
7.9	7.9	Send the client a pay less notice
7.12.2	7.12.2	Issue a final payment notice
8.1.1	8.1.1	Issue a notice of its intention to suspend some or all of its obligations
8.1.3	8.1.3	Suspend some or all of its obligations
9.3	9.3	Apply for a revision of time
9.7	9.7	Apply for an adjustment to the contract price
12.3	12.3	Issue a 14-day notice of intention to terminate
12.4	12.4	Terminate its employment by issuing a notice of termination
12.5	12.5	Terminate its employment
12.6	12.6	Terminate its employment
13.1	13.1	Attempt to settle disputes by mediation
13.2	13.2	Refer disputes to adjudication
13.8	13.8	If arbitration is selected, refer disputes to arbitration
13.9	13.9	If arbitration is not selected, refer disputes to the appropriate court

Management systems

3.36   The RIBA Building Contracts place more emphasis on management, and have more provisions concerning meetings, coordination and communication, than do other equivalent contracts.

3.37   These provisions are aimed at ensuring the smooth running of the project. They include requirements to:

- provide and update a programme;
- give early warning of matters that may cause a delay or affect the contract price (see paras 4.22 and 4.23);
- suggest improvements (see para. 6.8);
- hold a pre-start meeting;
- maintain a risk register, if appropriate.

Some of these mechanisms are discussed in other parts of this Guide, but the last two are of more general application.

Pre-start meeting

3.38   The requirement to hold a pre-start meeting, at least 10 days before the work is due to start on site, is contained in clause 3.1. Both parties must attend, as well as the contract administrator. Clause 3.1 states that the parties and the contract administrator must:

3.1.1 set out expectations from each other

3.1.2 set out the administration and communication procedures, including any specific rules on written and electronic communications

3.1.3 identify potential and actual risks and set out procedures to deal with them, to include the preparation of a Risk Register, if appropriate.

3.39   The pre-start meeting is often the point at which the contract documents are signed. None of the discussions ought to result in any changes to the contract terms or the obligations of the parties. However, it would be wise to bear in mind that there is nevertheless a close link between the above and various contractual issues. For example, the 'expectations' cannot be more than those that are already set out. The client, for example, cannot demand a higher standard of work than the level specified in the contract documents or that the contractor finishes earlier than the contractual date. If it becomes apparent at the meeting that expectations are at odds with those in the contract, then the documents should be amended before signature.

3.40   The meeting will be used to flesh out and fine-tune some of the procedural matters. Key among these would be whether the client is to remain in occupation, and what the exact arrangements will be with regard to security, safety, access, use of facilities, storage and disposal of rubbish. If any additional or new arrangements are made, these should be recorded and annexed to the contract documents. The meeting may also be used to agree communication procedures (cl. 3.1.2). Generally, all communication is to be in writing and issued to the parties and the contract administrator (cl. 11.8), unless the contract states otherwise. Special procedures are set out in the case of termination (cl. 11.9, see para. 9.17).

Progress meetings

3.41 The 2014 version of CBC required that progress meetings be held monthly (or other period as agreed) and attended by the contractor, the contract administrator and any other person invited by the contract administrator (CBC 2014, cl. 3.6). Although the requirement has been omitted, it is common practice to hold such meetings. Including the provision in CBC 2014 meant that it was an obligation on the contractor to attend, and a breach if it did not. If the parties feel that attendance at meetings is important, they may wish to consider adding in such a provision, either in the contract or within the preliminaries to the specification. The meetings will normally cover matters such as progress, technical issues and information needed, and are an opportunity to discuss advance warning notices. In particular, requiring the contractor to provide an update to its programme five days before the meeting would be a useful way for the contract administrator to monitor and record how the job is progressing.

Risk register

3.42 Clause 3.1.3 refers to the preparation of a risk register 'if appropriate' (in the 2014 editions this was included as an optional provision that, if selected, required the contract administrator to establish and maintain a risks register). The term 'Risk Register' is defined as 'a document identifying potential and actual risks which could affect the progress of the Works and setting out procedures to deal with them'. Initially, the register will list the risks and mitigation procedures identified and agreed at the pre-start meeting, but these are likely to be adjusted as the project progresses and new risks are identified. The register could be in the form of a simple list (or more likely a spreadsheet), and it is common practice to place a priority on the risks (e.g. highly likely, not likely), as well as to set out the measures or actions to be taken should a risk materialise and who will take them. Strictly speaking there is no contractual obligation to comply with any actions set out (unless, of course, the register includes obligations already covered in other contractual provisions). Nevertheless, it may be a useful point of reference in day-to-day communications and a good discussion item at progress and other meetings.

4 Project progress

4.1 Most building contracts include provisions that require the contractor to complete by a specific date or set of dates as this is often a matter of great importance to the client. Late completion may well result in losses to the client, in particular the costs of alternative accommodation and additional consultants' fees. However, the obligation to complete is normally subject to some exceptions; for example, the client will be required to allow the contractor additional time if the client itself causes delay. This chapter examines the contractor's obligations regarding time, programming and completion under the RIBA Building Contracts, including the mechanisms for monitoring progress and the sanctions for non-completion.

The site: possession

4.2 An address for the site is to be given in item C of the Contract Details. However, far more information may be needed than simply the address. If the contractor is not to have full access to the whole plot at that address – for example, if only part of the plot can be used, or if there are restrictions on entry points – this should be made clear at the tender stage. Similarly, if the building is to be occupied, the details of this should be given, including which parts and between which dates. Where the property shares common parts with other properties (e.g. stairs, parking), it may be sensible to explain exactly what the contractor may use of these.

4.3 The client has to allow access to the contractor for carrying out the works (cl. 1.2). Unlike some of the JCT contracts, the RIBA Building Contracts do not refer to the contractor having 'possession' of the site (which is normally held to be a licence to occupy the site up to the date of completion; *H.W. Nevill (Sunblest) Ltd* v *William Press & Son Ltd* and *Impresa Castelli SpA* v *Cola Holdings Ltd*). In the RIBA Building Contracts, the extent of the access will be interpreted in the light of all the information made available to the contractor at the time of tender. If little information is given, the courts will imply an obligation that the contractor should be given such possession, occupation or use as is necessary to enable it to perform the contract (*London Borough of Hounslow* v *Twickenham Gardens Development*). This may include access not just to the building where the work is to be carried out, but also to other areas in the control of the client (see *The Queen in Rights of Canada* v *Walter Cabbott Construction Ltd*).

> *The Queen in Rights of Canada* v *Walter Cabbott Construction Ltd* (1975) 21 BLR 42
>
> This Canadian case (Federal Court of Appeal) concerned work to construct a hatchery on a site (contract 1) where several other projects relating to ponds were also planned (contracts 3 and 4). The work to the ponds could not be undertaken without occupying part of the hatchery site. Work to the ponds was started in advance of contract 1, causing access problems to the contractor

when contract 1 began. The court confirmed (at page 52) the trial judge's view that 'the "site for the work" must, in the case of a completely new structure comprise not only the ground actually to be occupied by the completed structure but so much area around it as is within the control of the owner and is reasonably necessary for carrying out the work efficiently'.

Starting the work

4.4   The contractor is required to 'carry out and complete the Works in accordance with the Contract, in good and workmanlike manner, by the Date for Completion' (cl. 2.1.1). This wording has been amended slightly since the 2014 editions, which specifically required the contractor to start on the date stated in the contract.

4.5   It is often crucial that work actually starts on the defined date. In cases where the site is empty, the client needs to be assured that from that date the contractor will be responsible for security, health and safety, and general compliance with local authority requirements. An empty site is exposed and hazardous, so the client will be at risk of claims if security is not being addressed. Even where a client intends to remain in residence, it is disconcerting, and sometimes extremely inconvenient, if work does not commence on the planned date. Although there is no express requirement to begin work on the 'Start Date' specified in the contract details, it is likely that obligations regarding site security, etc, would run from that date. Clause 2.4, for example, requires that security is maintained 'at all times' and the contractor would have statutory obligations to ensure that the site is safe. However aside from this, there would be no obligation to begin building activity on the start date, provided that the work is finished by the completion date; if an early start is important then additional terms may need to be added.

4.6   There are provisions to allow the client to defer access to the whole site, or to 'Sections of the Site', if needed (cl. 1.4). This can be very helpful if, for example, the client has problems arranging alternative accommodation, or for removal of furniture or equipment to storage. The contract does not place any limits on the length of deferral allowed, nor does it require the client to give any advance warning notice. However, the contractor would be entitled to a revision of time, and to additional payment to cover costs arising as a result of the delay. As these could be considerable, especially if little or no warning is given, the client would be wise to exercise the right to defer only in emergency situations, or if the deferral and the contractual implications can be agreed with the contractor well in advance. It should be noted that the term 'Sections' refers to sections that are pre-defined in the Contract Details (see below), so that in theory clause 1.4 does not allow the client to defer access to parts of the site on an ad hoc basis without prior agreement of the contractor. However, as the contractor will be compensated for any losses, in practice it is unlikely to object.

Completion in sections

4.7   The RIBA Building Contracts include an optional clause (cl. 17) providing for completion in sections. Although the clause refers only to 'Completion', it is possible to arrange for the work to be started and/or finished in sections. This would be useful where the client cannot make all areas of the site available at the same time, or needs certain parts of the building before others. On large projects, phased working is quite often more economical for the contractor, as it can move its resources around the various sections in a phased programme. On smaller projects, however, it may actually be less convenient, so this may need to be negotiated following tender submissions. If this option is selected, separate start and completion dates and rates of liquidated damages are specified for each section in item R of the Contract Details.

The contractor's programme

4.8 Both versions of the contract contain an optional clause (cl. 14), whereby the contractor is required to provide a programme. It is recommended that this should always be selected (item O), except on the very smallest of projects. The programme is of great help for:

- giving the client a general idea as to what to expect (especially important to the client if they are in residence);
- giving the contract administrator an indication as to when the contractor will require further information;
- alerting the contract administrator as to when it may be wise to inspect the site;
- acting as an early alert if the contractor is slipping behind programme or getting into financial difficulties.

4.9 The programme is not identified as being one of the contract documents under item F of the Contract Details. Nevertheless, it would be sensible to ask that the contractor provides a programme before the contract is entered into. This could be done either by requiring tenderers to provide a programme with their tenders, or, once the tenders have been received, by requesting the preferred bidder to submit a programme before its tender is accepted. Otherwise, the programme is to be provided no later than the date of the pre-start meeting (cl. 14.1).

4.10 There are no sanctions for non-production of a programme, therefore it would be sensible to insist on it before the contract is executed. The 2014 editions contained some alternative sanctions, which were a useful provision (it is not clear why these were removed). Under these, the contractor could either be subject to a financial penalty or prevented from starting work until the programme is produced. The contractor was still bound to finish by the completion date, despite the delayed start. The financial penalty involved withholding 10 per cent of the value of the first payment certificate until the programme was produced. Either option would be effective, but preventing start on site would offer a more stringent sanction. If required, the contract could be amended to add one of these sanctions back in.

Content of programme

4.11 The parties are required to set out some information in the Contract Details, under item O, namely: 'the activities the Contractor will carry out to complete the Works, the start and finish dates of each activity and the relationship of each activity to the others, which may include lead and lag times'. In the 2014 version, this was stated to be something that the contractor should include in its programme; in the 2018 version the information must actually be inserted in the Contract Details. This could only be done, of course, if this has already been provided – it appears to assume that the contractor's programme will have been submitted before the contract is executed (as noted above, this is good practice).

4.12   This information will, among other things, give a clear idea of the sequence of work operations throughout the project. Including 'the relationship of each activity to the others' will show whether an item needs to be finished before another can start, i.e. which items are time critical. The shortest route through all time-critical activities is usually referred to as the 'critical path'. Knowing the critical path is extremely useful when it comes to assessing revisions of time. It should be noted that the critical path is not fixed and may change throughout the project, therefore having regular programme updates is essential.

4.13   There are other matters than can be very useful to know, for example the number of people and other resources. This gives an immediate indicator of whether the contractor is not resourcing the project as planned, and therefore may be evidence that the contractor is responsible for a delay. In extreme cases it may signal that the contractor is getting into financial difficulty.

4.14   Exactly what information is required would need to be set out at tender stage, otherwise there would be no obligation on the contractor to provide it. These days most contractors use software packages to work out their programmes, and the packages would always show both sets of information (in fact, an input of resources is required in order for the package to calculate the durations of activities).

Drawings/information required/provided

4.15   One of the further requirements for the programme might be to show the dates when the contractor will require additional information or drawings. There is no requirement in the contracts for further information to be provided, but it is likely that such a duty would be implied (see para. 5.10). It would therefore be useful for the contract administrator to be aware of when the contractor anticipates it will need information. Nevertheless, as the programme is not a contract document, any dates shown would not be binding.

Progress

4.16   The contractor is required to 'carry out and complete the Works in accordance with the Contract' (cl. 2.1.1). Unlike the equivalent clause in CBC/DBC 2014, this clause no longer contains an express provision to proceed 'regularly and diligently'. However it should be noted that cl 12.1.2 gives failure by the contractor to proceed regularly or diligently as a ground for termination by the client (see paras 9.6 and 9.8–9.11), therefore the contractor must maintain reasonable progress. The contractor does not have to stick precisely to its own programme, so long as it completes by the date for completion; if the client requires any parts to be completed before others, or to use any parts during the course of the works, it will need to make use of the sectional completion provisions, or those for partial possession before practical completion (see paras 7.9–7.12).

4.17   The contract administrator will not normally intervene in matters concerning the day-to-day programming. As the contract administrator is given the specific power to issue instructions 'postponing the Works or one or more Sections of the Works' (cl. 5.4.2) it would be entitled to alter the working sequence, but should generally only use this power where no other option is open. Examples might be if the local authority requires work to cease for a period of time, if an unanticipated health and safety hazard is encountered, or if the client has second thoughts about the design of a part of the project and work has to be put on hold while discussions are held. There will almost always be significant effects on the completion date and the costs of the project.

Updated programmes

4.18   The 2014 edition of CBC (but not DBC) included a useful provision which required the contractor to submit regular programme updates. As published, the 2018 no longer requires the contractor to do this, even following revisions to the contract period (as it would in JCT contracts). In most projects, the original programme is amended many times, even when the overall completion date is met, and certainly if there are major delays or an extension is awarded. As such, the original programme is usually of little use by the time the project is even halfway through. The parties may wish to consider adding a requirement for a monthly update to be provided.

4.19   In CBC 2014, the updates were linked to progress meetings (also omitted), and the contractor was required to submit an updated programme to the contract administrator no later than 5 days before each progress meeting, and a revised version shortly after if amendments were agreed. This is the system used in the Government General Conditions of Contract for Building and Civil Engineering (GC Works) contracts, and which has proved to be very effective in practice (it is also used in NEC4). For clarity, the 'update' was to be provided even if there were no changes to the programme. Even without any delays being experienced, the contractor will in practice regularly review and make adjustments to the resourcing and sequencing, and it is very useful for the contract administrator to be made aware of these.

Finishing the work

4.20   The Contract Details require the insertion of a 'Date for Completion' (and if completion in sections is selected, a date for completion for each section), and the contractor is required to complete by this date (cl. 2.1.1). However, this is rarely the date the works are actually completed. If delays occur that were not the fault of the contractor, the contractor will be entitled to a 'Revision of Time'; this will mean that the date for completion of the works or a section will be postponed, and the contractor's obligation will be to meet the revised date. If the contractor simply fails to finish by the original or revised date for completion, liquidated damages will be claimable by the client (note: the damages may need to be repaid if the date for completion is subsequently revised).

4.21   Once the works are complete, the contract administrator will certify 'Practical Completion', which is then followed by a 'Defects Fixing Period'. When all defects have been corrected, the contract administrator will issue a final payment certificate. The rest of this chapter considers delays to the programme. Practical completion and the remedying of defects are dealt with in Chapter 7.

Delay

4.22   Delays to a project will obviously cause problems for all involved, including the client, the contractor and the contract administrator, and so should be prevented or minimised if at all possible. The 'Collaborative Working' section that appears in both of the RIBA Building

Contracts includes provisions that are aimed at preventing or managing delay and its consequences, under the heading 'Advance Warning and Joint Resolution of Delay'. The relevant clauses state:

3.2   If an event occurs which affects or is likely to affect the progress of the Works and/ or the Contract Price, the Parties shall:

3.2.1   notify each other and the Architect/Contract Administrator of the event as soon as they become aware of it

3.2.2   work together to resolve the event. If necessary the Architect/Contract Administrator may hold a meeting with the Parties and other related stakeholders to do this

3.2.3   take reasonable steps to minimize the effects of the event on the Contract.

4.23   The obligation applies to both parties; so, for example, the client might notify of a possible restriction to access, and the contractor might notify of a problem with obtaining a material. A notification by the contractor must be given whether or not the event might entitle it to a revision of time. This requirement is repeated in clause 9.4, where the contractor must inform the contract administrator of 'any event affecting the progress of the Works'.

4.24   There is no obligation on the contract administrator to issue a notification of events of which only it is aware, but of course it would be sensible for it to do so. Also, although the clause states 'affects or is likely to affect', it would be sensible for a party to alert the other to events that might occur as well as ones that will occur (if there is a reasonable probability). The aim of the warning is to allow a strategy to resolve or minimise the effect of the event to be agreed in advance of the event occurring, and some events might be entirely averted if a warning is issued sufficiently early. Whether or not a warning is given, the contractor is required to take reasonable steps to minimise the effect of any delaying event (cl. 3.2.3).

Revisions to the contract completion date

4.25   All construction contracts include provisions for making revisions to the contract period (usually referred to as 'extensions of time', but in the RIBA Building Contracts they are referred to as 'Revisions of Time'). In many cases, construction contracts will list the reasons that would justify a revision, which will comprise 'neutral' events (i.e. things that could occur through neither party's fault, such as bad weather) and events caused by actions of the client or the contract administrator. Such a list of events acts as a means of distributing risk between the parties; the more events are included, the more risk is borne by the client.

4.26   The main reason for including events caused by the client and the contract administrator is to preserve the client's right to liquidated damages. If no such provisions were included and a delay occurred that was caused by the client, this would in effect be a breach of contract by the client and the contractor would no longer be bound to complete by the completion date (see *Peak Construction* v *McKinney Foundations*). The client would therefore lose the right to liquidated damages, even if some of the blame for the delay rests with the contractor. The phrase 'time at large' is often used to describe this situation. However, this is, strictly speaking, a misuse of the phrase as in most cases the contractor would remain under an obligation to complete within a reasonable time.

Peak Construction (Liverpool) Ltd v *McKinney Foundations Ltd* (1970) 1 BLR 111 (CA)

Peak Construction was the main contractor on a project to construct a multi-storey block of flats for Liverpool Corporation. The main contract was not on any of the standard forms, but was drawn up by the Corporation. McKinney Foundations Ltd was the subcontractor nominated to design and construct the piling. After the piling was complete and the subcontractor had left the site, serious defects were discovered in one of the piles and, following further investigation, minor defects were found in several other piles. Work was halted while the best strategy for remedial work was debated between the parties. The city surveyor did not accept the initial remedial proposals, and it was agreed that an independent engineer would prepare an alternative proposal. The Corporation refused to agree to accept his decision in advance, and delayed making the appointment. Altogether it was 58 weeks before work resumed (although the remedial work took only six weeks) and the main contractors brought a claim against the subcontractor for damages. The Official Referee, at first instance, found that the entire 58 weeks constituted delay caused by the nominated subcontractor and awarded £40,000 damages for breach of contract, based in part on liquidated damages which the Corporation had claimed from the contractor. McKinney appealed, and the Court of Appeal found that the 58-week delay could not possibly entirely be due to the subcontractor's breach, but was in part caused by the tardiness of the Corporation. This being the case, and as there were no provisions in the contract for extending time for delay on the part of the local authority, it lost its right to claim liquidated damages, and this component of the damages awarded against the subcontractor was disallowed. Even if the contract had contained such a provision, the failure of the architect to exercise it would have prevented the Corporation from claiming liquidated damages. The only remedy would have been for the Corporation to prove what damages it had suffered as a result of the breach.

4.27   In both CBC and DBC, the provisions are set out in clause 9.3, which states:

> 9.3   The Contractor may apply (with supporting documentation) for a Revision of Time if the Works are or are likely to be delayed by any of the following:
>
> 9.3.1   the Architect/Contract Administrator issues a Change to Works instruction
>
> 9.3.2   the Client defers or withdraws access to the Site
>
> 9.3.3   the Client or its agents cause delay or disruption to the Works or part of the Works
>
> 9.3.4   the Client or its agents cause the Works or part of the Works to be suspended
>
> 9.3.5   subject to clause 5.5.1, an instruction issued for work to be uncovered and inspected/tested under clause 5.5
>
> 9.3.6   the action or omission of a utility company or statutory body, subject to advance warning notification under clauses 3.2 and 3.3
>
> 9.3.7   any event under clause 8 (Contractor's Right to Suspend)
>
> 9.3.8   any event under clause 9.1 (Force Majeure).

4.28   Many of these events are dealt with at other points in this Guide: for change to works instructions, see paragraphs 5.23–5.31; for deferring access and postponing work, see paragraphs 4.6 and 4.17 above; for inspections and tests, see paragraphs 5.19–5.22; and for inconsistencies, see paragraphs 2.44 and 2.45. In regard to other events, the following points should be noted:

- The client or its agents causing delay or disruption (cl. 9.3.3) acts as a useful 'catch-all', to pick up any actions not covered elsewhere in the list. 'Agents' would include the contract administrator if acting on behalf of the client, and may include other third parties whose actions have been authorised by the client.

- 'Force Majeure' is a defined term, and covers matters for which the client has agreed to accept the risks of any delay. It is more widely defined than the normal understanding in law.

4.29   There is no reference to delays due to bad weather – any application for delays due to weather would need to show that it fell within the definition of force majeure, in which case the weather event would have to be exceptionally bad, and not one that could be expected at the time of year in question. Any ambiguity here can usually be resolved by consulting the contractor's site records and/or Meteorological Office records.

Applying for a revision of time

4.30   If the contractor is delayed by a clause 9.3 event and wishes to apply for a revision of time, it must do so, with supporting information 'within a reasonable time of its occurrence'. The contract administrator and contractor are required to 'aim to agree the Revision of Time promptly' (cl. 9.5). If no agreement can be reached, the contract administrator is required to 'make a reasonable assessment of the Revision of Time, taking into consideration all the supporting documentation' (cl. 9.6). Although the clause states 'may', the contractor administrator should make a decision within a reasonable period, so that the parties understand the position. Note that the clauses overlap to a certain extent those dealing with change to works instructions. These essentially follow the same sequence, except that if the contractor wishes to make an application it should do so within 7 days of receiving the instruction.

4.31   Surprisingly, having decided a revision is appropriate, there appears to be no obligation on the contract administrator to amend the date for completion, or to inform both parties, although obviously it would be sensible to do both.

4.32   It not clear whether the contractor's application is a condition precedent to the award of a revision of time, i.e. not only would it lose the right, but the contract administrator would have no power to issue a revision unless the contractor has made an application. It is unlikely that the contractor would object to a revision made by the contract administrator on its own initiative, and therefore unlikely that any such revision would cause problems. Nevertheless, it would be wise for the contract administrator to seek the agreement of both parties before making the revision.

Assessment of an application

4.33   Assessment of revisions of time is a complex process that often causes difficulty in practice. What follows is a brief outline only. If faced with difficult claims, the contract administrator should consult one of the published texts on the subject (e.g. Birkby et al., 2008; Eggleston, 2009) or take expert advice.

4.34 All assessments must be made in a fair and reasonable manner. If the parties have agreed to use any rules, such as the SCL Delay and Disruption Protocol, then these should be taken into account. The objective is always to assess what effect the event will have on the final completion date, and the contract administrator should take into account the fact that the contractor should use reasonable endeavours to minimise the effects of any delaying event. This would include giving an early warning of any event of which it ought to be aware, and taking reasonable steps to reorganise the works and adjust the programme (cl. 3.2.3).

4.35 It should be remembered that the standard of proof is 'on the balance of probability' (the civil standard), i.e. the contractor has to convince the contract administrator that it is more likely than not that it suffered delay due to the event. The contract administrator should not expect the application to prove the contractor's case 'beyond all reasonable doubt' (the criminal standard). The contractor must demonstrate not only that an event listed in clause 9.3 has occurred, but also that the event has delayed the work, and that the particular work delayed is on the critical path, i.e. its delay will ultimately delay the completion of the project; put simply, it must show a causal link between the event and delay to completion.

4.36 The contract administrator is required to take into account any supporting documentation when making adjustments to the time (cl. 9.6). This could include, for example, the contractor's programme. The contract administrator could also take into account any advance warning notices. There are probably two ways in which the notices may be relevant: first, the fact that they were given is evidence that the contractor has been vigilant and used reasonable endeavours to avoid the delay (conversely, if no warning is given this must be taken into account, cl. 3.3); and second, the notices should contain useful contemporaneous evidence of the nature and anticipated effects of the event. The contract administrator could also consult any other available records as to the history of events on site, and use its own knowledge and experience when making the decision.

4.37 Two issues in particular can cause problems when assessing revisions of time: 'concurrent delay' and 'contractor's float'. The following sections present very brief explanations of these issues.

Concurrent delay

4.38 Where two separate events contribute to the same period of delay, but only one of these is an event listed in clause 9.3, the normal approach is that the contractor is given a revision of time for the full effect of the clause 9.3 event (i.e. the contractor gets the benefit of the contributing but approximately equal cause, unless another competing cause can be identified as the dominant cause). The courts have normally adopted this approach (see *Walter Lilly & Co Ltd* v *Giles Mackay & DMW Ltd*; note, however, that the same does not apply to claims for loss/expense).

4.39 The instinctive reaction of many assessors might be to 'split the difference', given that both parties have contributed to the delay. However, it is more logical that the contractor should be given a revision of time for the full length of delay caused by the relevant event, irrespective of the fact that, during the overlap, the contractor was also causing delay. Taking any other approach, e.g. splitting the overlap period and awarding only half of the extension to the contractor, could result in the contractor being subject to liquidated damages for a delay partly caused by the client.

Walter Lilly & Co. Ltd v *Giles Mackay & DMW Ltd* [2012] EWHC 649 (TCC)

This case concerned a contract to build Mr and Mrs Mackay's, and two other families', luxury new homes in South Kensington, London. The contract was entered into in 2004 on the JCT Standard Form of Building Contract 1998 Edition with a Contractor's Designed Portion Supplement. The total contract sum was £15.3 million, the date for completion was 23 January 2006, and liquidated damages were set at £6,400 per day. Practical completion was certified on 7 July 2008. The contractor (Walter Lilly) issued 234 notices of delay and requests for extensions of time, of which fewer than a quarter were answered. The contractor brought a claim for, among other things, an additional extension of time. The court awarded a full extension up to the date of practical completion. It took the opportunity to review approaches to dealing with concurrent delay, including that in the case of *Henry Boot Construction (UK) Ltd* v *Malmaison Hotel (Manchester) Ltd* (where the contractor is entitled to a full extension of time for delay caused by two or more events, provided one is an event which entitles it to an extension under the contract), and the alternative approach in the Scottish case of *City Inn Ltd* v *Shepherd Construction Ltd* (where the delay is apportioned between the events). The court decided that the former was the correct approach in this case. As part of its reasoning the court noted that there was nothing in the relevant clauses to suggest that the extension of time should be reduced if the contractor was partly to blame for the delay.

Float

4.40   Another area that sometimes causes problems is the question of float. Float is essentially planned early completion, i.e. a period shown on a programme between the contractor's planned completion date and the contractual date for completion. If a revision of time is applied for at a relatively early stage in the project, it may be that the delay suffered will not push the planned date beyond the contractual date for completion. Therefore, strictly speaking, no revision should be given. However, if the contractor is later delayed through its own errors, it may wish it had had the benefit of the earlier revision, as it now appears unlikely that it will complete on time. In such cases it is generally considered that the contractor should be given the benefit of the 'float', therefore the contract administrator may need to review earlier decisions and account for the float period.

Final assessment of revisions of time

4.41   The RIBA Building Contracts do not require the contract administrator to review any revisions of time after practical completion. However, the RIBA has confirmed that it would be possible to do this, provided that it extended, not reduced, the contract period. After practical completion, the contract administrator will be able to review all the earlier revisions of time and to adjust the date for completion as necessary with the benefit of full information, including the final programme. Such an adjustment would only be to extend the date further; the contract administrator would not be entitled to shorten the programme at that stage.

5 Control of the works

5.1   The previous chapter focused on the programming of the works, and on monitoring progress in relation to the planned programme. This chapter examines the quality of the works: how it is achieved, who is responsible for it and what steps can and may need to be taken if problems are experienced.

5.2   Achieving the contractual standard is entirely the responsibility of the contractor, but the contract administrator also has a role to play, by providing information on the required standard and monitoring whether it is achieved. The RIBA Building Contracts also confer various powers on the contract administrator that can be used if the contract administrator feels it is necessary to step in.

Control of day-to-day activities

5.3   The day-to-day control of the works, i.e. the management of operations on site, coordination of orders and supplies, procurement of labour and subcontractors, and all issues relating to quality control, is entirely the responsibility of the contractor.

Contractor's representative

5.4   In order that its responsibility is carried out properly, the contractor is required to ensure that 'a suitably qualified representative is available during the Works to answer queries and receive instructions on its behalf' (cl. 2.3). What constitutes 'suitably qualified' would depend on the nature and scale of the project, i.e. the representative should be sufficiently qualified to fulfil their role competently and in accordance with the contract. The contracts do not require that the person is present on site or available 'at all times' (as in the JCT's SBC16), and it may be that something less than full-time presence would be acceptable, provided that they are available at all material times, and that they make arrangements for dealing with queries or receiving instructions during short absences. There is no requirement in the Contract Conditions to have the person named, but it would be good practice to establish the identity of the person at the pre-start meeting, and to make sure this is recorded in writing.

Responsibility for subcontractors

5.5   The contractor is also fully responsible for the quality of work of all subcontractors (cl. 2.6), whether these are its own domestic subcontractors or those selected by the client under the required specialists optional clause 16. As there are no provisions in the RIBA Building Contracts for the client to directly engage other firms, any persons on the site during the works would be under the direct supervision and responsibility of the contractor. If the client made any special arrangements with the contractor, outside of the contract, to allow its directly engaged persons on site, then this would cause confusion as to who is responsible for their performance, including in relation to quality of work, progress and health and safety, unless a detailed agreement regarding these matters is drawn up.

Principal designer

5.6 The RIBA Building Contracts require that both parties comply with all health and safety regulations (cl. 1.5 and 2.1.3). The key relevant regulations are the Construction (Design and Management) Regulations 2015, which came into force on 6 April 2015 and apply to all construction projects.

5.7 For almost all projects, the Regulations require the client to appoint (in writing) a principal designer and a principal contractor (Regulation 5). The principal designer manages and coordinates health and safety aspects during the pre-construction phase, and then liaises with the principal contractor and coordinates ongoing design work during the construction phase. The principal designer could be the architect and/or the contractor administrator, but this is not necessarily the case; the role of principal designer is a distinct one and should normally be covered by a separate appointment. The principal contractor, which will almost always be the contractor under the contract, manages the construction phase of a project. This involves liaising with the client and principal designer throughout the project, including during the pre-construction phase, and producing a plan of how it will manage health and safety on site during the construction phase. If a domestic client fails to make the required appointments, the Regulations state that the designer in control of the pre-construction phase of the project is the principal designer, and that the contractor in control of the construction phase is the principal contractor (Regulation 7(2)).

5.8 If either party breaches its obligations under clause 1.5 or 2.1.3, it will be contractually liable to the other party, as well as liable under the Regulations. For a detailed understanding of their roles, the parties should consult the Regulations and related guidance.[1]

Flow of information

5.9 The most important function of the contract administrator is to ensure that the contractor is supplied with detailed and accurate information, either at tender stage or during the project, which makes clear precisely what standards and quality are required. In projects where the contractor is undertaking design, it will be required to submit its design to the contract administrator before or during the construction phase. The overall responsibility for integrating that design with the rest of the project rests with the contract administrator.

Information to be provided by the contract administrator

5.10 In most projects the information in the contract documents is not sufficient to construct the works; a certain amount of detailed information (e.g. schedules of finishes) is often outstanding, and it is usual for the contractor to be provided with this information during

---

[1]   See, for example, the Health and Safety Executive Guidance on Regulations L153 *Managing Health and Safety in Construction*, and the CIC Risk Management Briefing *Construction (Design and Management) Regulations 2015*.

the course of the work. The RIBA Building Contracts state at clause 5.2 that the Contract Administrator:

> 5.2.1 provides the Contractor with up to two copies of the Contract Documents
>
> 5.2.2 issues any required changes and variations.

5.11   Although this does not refer to additional information, only to 'required changes and variations', it is likely that it would be given a broad interpretation that would include any information necessary to complete the project. In any case, such an obligation is likely to be implied with respect to all work that is not listed as a part to be designed by the contractor. It is difficult to see how a contractor could successfully achieve completion in the absence of a duty requiring that instructions, information, plans, drawings etc. are issued in good time, so the obligation is likely to arise under an implied duty to cooperate (see *Wells* v *Army & Navy Co-operative Society Ltd*, also para. 3.15, and *National Museums and Galleries on Merseyside* v *AEW Architects and Designers Ltd*). The contract administrator should therefore assume that it should provide the contractor with all key information, the only exceptions being in relation to contractor-designed items and, possibly, very small items, where it may be that the contractor could be expected to determine these for itself.

> *Wells* v *Army & Navy Co-operative Society Ltd* (1902) 86 LT 764
>
> The Court of Appeal refused to allow the deduction of liquidated damages where late completion was partly caused by late provision of information by the architect (as well as by variations and a delay by the client in giving possession). These were considered acts of prevention that were not catered for in the extension of time clause in the contract, and it was held that the liquidated damages provisions were ineffective and could not be applied to the delay period.

5.12   The time by which information should be supplied is whenever the contractor needs it, given the overall progress on site and the date for completion. The contractor's programme might have been required to set out dates for information to be provided, in which case this would be a guide, but it would not be conclusive. As with other issues concerning timing, it would be sensible to have this as an ongoing agenda item at progress meetings.

Information to be provided by the contractor

5.13   If optional clause 15 is selected, the contractor is required to provide information regarding its developing design, as set out in the Contract Details. Clause 15.2 states:

> At least 21 days before carrying out any part of the Works listed under item P of the Contract Details, the Contractor shall submit details of its design to the Architect/ Contract Administrator for comment.

This is a very useful provision to include, as it is essential that the contract administrator is able to monitor that the design meets the client's requirements, as set out in the contract documents, and to coordinate it with the rest of the project.

5.14   The RIBA Building Contracts do not include a detailed procedure for submissions (e.g. format, response, comments or re-submission), such as that set out in Schedule 1 to

SBC16. The administrator could, however, use the power under clause 5.4.5 to instruct that further or revised documents are provided. More importantly, the contracts do not include an equivalent clause to SBC16 Schedule 1:8.3, which states that:

> neither compliance with the design submission procedure in this Schedule nor with the Architect/Contract Administrator's comments shall diminish the Contractor's obligations to ensure that the Contractor's Design Documents and CDP Works are in accordance with this Contract.

5.15   If an RIBA Building Contract is to be used for a project where there are significant contractor design elements, then it may be wise to consider including some provisions regarding submission. Failing that, when making comments it might be sensible to remind the contractor of its obligation to ensure that the design is in accordance with the client's specification (cl. 15.1.2).

Inspection and tests

5.16   In addition to providing information, the contract administrator will also inspect the works at regular intervals to monitor whether the required standard is being met. The contract administrator may also, if necessary, issue instructions to have work opened up and tested, although this may have implications for the contract price and programme.

Inspection

5.17   On most projects the contract administrator will inspect the works at regular intervals. The RIBA Building Contracts do not place a duty on the contract administrator to do this, although they do give the contract administrator the power to visit the site and other off-site locations, and inspect the works (cl. 5.3). Note that the reference in the guidance notes to the contract administrator's 'duty' to visit is incorrect. However, the contract administrator's obligations to the client will, almost always, include a duty to inspect. Normally this would be an express duty under the terms of appointment, but in some circumstances it could also be implied: clearly, when the contract administrator is required under the contract to form an opinion on various matters – including being satisfied with the standard of work and materials prior to issuing a payment certificate – then it is essential that some form of inspection takes place. However, it is important to note that the duty is owed to the client, and not to the contractor. For example, a contractor cannot blame a contract administrator for failing to draw its attention to defective work.

5.18   Furthermore, a contract administrator will not necessarily be liable to the client for negligent inspection if a defect in a contractor's work is not identified. The question in every case is whether the contract administrator exhibited the degree of skill that an ordinary competent professional would exhibit in the same circumstances. Generally, the extent and frequency of inspections must enable the contract administrator to be in a position to properly certify that the construction work has been carried out in accordance with the contract (*Jameson* v *Simon*). The case of *McGlinn* v *Waltham Contractors Ltd* sets out some useful advice on the appropriate standard of inspection.

McGlinn v *Waltham Contractors Ltd* [2007] 111 Con LR 1

This case concerned a house in Jersey called '*Maison d'Or*' that was designed and built for the claimant, Mr McGlinn. The house took three years to build, but after it was substantially complete, it sat empty for the next three years while defects were investigated. It was completely demolished in 2005 having never been lived in, and was not rebuilt. Mr McGlinn brought an action against the various consultants, including the architect, and the contractor, claiming that *Maison d'Or* was so badly designed, and so badly built, that he was entitled to demolish it and start again. The contractor however had gone into administration and played no part in the hearing. The architect was engaged on RIBA Standard Form of Appointment 1982, which referred to 'periodic inspections'. HH Judge Peter Coulson QC usefully summarised the principles relating to inspection (at paras 215 and 218), which included the following:

- The change from 'supervision' to 'inspection' represented 'a potentially important reduction in the scope of an architect's services'.

- 'The frequency and duration of inspections should be tailored to the nature of the works going on at site from time to time.'

- 'If the element of the work is important because it is going to be repeated throughout one significant part of the building, then the inspecting professional should ensure that he has seen that element of the work in the early course of construction/assembly so as to form a view as to the contractor's ability to carry out that particular task.'

Testing and defective work

5.19 As noted at the beginning of this chapter, it is entirely the contractor's responsibility to ensure that the work is completed in accordance with the contract. The contract administrator is, however, given various discretionary powers that may be useful if it is concerned that the contractor does not appear to be fulfilling this primary obligation.

5.20 First, the contract administrator may issue instructions requiring any work to be uncovered, inspected and/or tested for compliance (cl. 5.5). If the work proves to be defective, the contractor will bear the cost of complying with the instruction and the correction of the defects (cl. 5.5.1). If the work complies with the contract, the client will bear the cost of complying with the instruction (cl. 5.5.2). Generally, therefore, the contract administrator would only issue such an instruction if there was a serious concern, or if the failure of the element in question would be crucial to the project or extremely difficult to correct later. Failure to issue any instructions would not in any circumstances lessen the contractor's responsibility, no matter how difficult or expensive it might be to correct the problem later.

5.21 Second, whether or not the defective work has been tested, the contract administrator has the power to reject work that is not in accordance with the contract (cl. 5.4.4).

5.22 Unlike the 2014 versions, the current contracts do not contain an express power whereby the contract administrator may accept work that does not accord with the contract 'and adjust the Contract Price accordingly' (formerly cl. 5.8 in CBC, or cl. 5.5.3 in DBC). However, if necessary this could be done by issuing an instruction implementing a change under clause 5.4.1. Care should be taken when doing this. The contract administrator should obtain the client's agreement, and a value should be proposed and agreed (see para. 6.11), and it would be advisable for the contract administrator to confirm the agreed deduction in the instruction.

Contractor administrator's instructions

5.23   The RIBA Building Contracts give the contract administrator the power to issue a range of instructions. In some cases these are expressed as being a duty, and generally if the contract administrator fails to issue instructions necessary for the progress of the works, this may constitute a breach by the client.

5.24   Unlike the 2014 versions, there is no general clause that gives the contract administrator a wide discretionary power ('instructions on any clause of the Contract to enable good administration'). Although clause 5.1 states 'The Architect/Contract Administrator is not a Party to the Contract but administers the Contract, issuing instructions and certificates and taking decisions' this is effectively a description of the role, not a clause conferring powers in addition to those detailed elsewhere. The contract administrator should therefore be careful that any instruction is empowered under a specific provision of the contract, and it would be good practice to state the relevant clause in the instruction.

5.25   A key power of the contract administrator is to issue an instruction that requires a change to the works (cl. 5.4.1). Note that a 'change' is not defined, nor is the exact extent of this power, as it is for example in JCT contracts, where it covers not only the design of the works, but also the manner of carrying out the works, working hours, access to the site and general management matters. The contract administrator is, however, given the power to postpone the works (cl. 5.4.2).

Delivery of instructions

5.26   All instructions are required to be in writing (cl. 11.8), and the contractor is required to comply with them promptly (cl. 5.6). If the contract administrator gives an instruction otherwise than in accordance with clause 11.8, for example orally, the contract requires that it is confirmed in writing 'promptly' (cl. 5.7).

5.27   The contracts do not make reference to the common practice whereby the contractor issues a 'written record of the oral instruction' (often termed a 'confirmation of verbal instruction' or 'CVI'). Such a confirmation would therefore have no contractual effect, although it may serve as a useful reminder to the contract administrator. The contract administrator, however, should avoid slipping into a pattern where it relies entirely on the contractor's CVIs as a trigger to issuing necessary instructions, and must be very vigilant and check that a contractor's confirmation exactly reflects what was intended. In a busy office with a constant stream of emails, it is easy to misread one, and an inaccurate confirmation could inadvertently be turned into a binding instruction.

5.28   Ideally, contract administrators should avoid giving oral instructions, except in cases of emergency. If they cannot be avoided, it should not be difficult to confirm an instruction promptly using electronic communications. Remember that a simple email would constitute an instruction in writing; there is no need (although it may be good practice) to issue it in any special format.

Procedure following an instruction

5.29   The contractor is required to notify the contract administrator, within 7 days of receiving an instruction requiring a change to the works, if it believes the instruction is not in accordance with the contract or that implementing it would have adverse health and safety implications or would adversely affect any part of the works designed by the contractor (cl. 5.14). The contract administrator may then, on receipt of the notification, modify, amend, withdraw or confirm the instruction, and the contractor shall comply accordingly (cl. 5.15).

5.30   In both contracts, if the contractor fails to comply with an instruction, the contract administrator may issue the contractor with a 7-day notice to comply (cl. 5.8), and if the contractor fails to comply with the notice, the client may engage others to undertake the instruction (cl. 5.8.1); this works in a very similar way to the notice to comply with provisions in JCT contracts. As with the JCT notice, if the contract administrator has serious concerns that the contractor may refuse to comply (e.g. because it has already expressed that intention at a meeting) then it would be possible for it to issue the instruction and the compliance notice together. In addition to giving the client this right, the contracts require the contractor to cooperate with the new contractors (cl. 5.8.2) and to allow them access to the site (cl. 5.8.3), and the contractor is to be responsible for all costs and expenses incurred by the client (cl. 5.8.4).

5.31   Where a change to works instruction is issued, the contractor is required to calculate and submit details to the contract administrator of its effect on the contract price and date for completion (cl. 5.11). The contract administrator and contractor should aim to agree the appropriate revision of time and/or additional payment promptly (cl. 5.12), otherwise the contract administrator determines the appropriate amount (cl. 5.13).

6 Payment and certification

6.1 In any building contract, the key obligation of the client is to ensure the contractor is paid according to the contract, both in terms of the amounts paid and the timing of these payments. To pay the contractor less than it is rightfully due or to pay the right amounts but later than agreed will put financial strain on the contractor (normally through increased borrowing costs), which in extreme cases could result in bankruptcy. On the other hand, paying too much, or too early, will put the client at risk; should the contractor repudiate the contract it will be difficult, and in some cases impossible, to recover the money.

6.2 The RIBA Building Contracts contain detailed provisions concerning the appropriate amounts due to the contractor, how and when these amounts are to be assessed, when they become due and the procedures for payment, all of which are discussed below.

6.3 Both contracts offer a choice between monthly certification based on the value of work completed, a single interim payment following practical completion, or milestone payments. In addition, CBC offers the option of making an advance payment.

The contract price

6.4 The 'Contract Price' is defined in both contracts as 'the amount that the Client shall pay the Contractor for carrying out and completing the Works, calculated in accordance with clause 7 of the Contract'. The Agreement states 'The Client shall pay the Contractor the Contract Price, which will be calculated in accordance with the Contract'.

6.5 The Contract Price is set out in item K of the Contract Details of both contracts. This can be either a lump sum (the sum is inserted) or an amount calculated in accordance with the 'Pricing Document listed under item D'. This would normally be a schedule of rates, in which case the amount due will be calculated in relation to the quantity of work actually carried out, based on the schedules of rates and prices provided by the contractor at tender. However, even if a lump sum, the contract price will be subject to change. This is acknowledged under the definition of 'Contract Price', which notes 'The Contract Price may increase (or decrease) as a result of instructions given by the Architect/Contract Administrator'. In fact there are a range of mechanisms whereby the contract price may be adjusted, which are discussed in the next section.

Adjustments to the contract price

6.6 References to adjustments to the contract price can be found in the following clauses:

- (cl. 3.4) contractor's proposed changes that will improve quality and/or reduce the contract price;

- (cl. 5.5.2) costs resulting from instructions regarding tests, etc. being added to the contract price;
- (cl. 5.11) adjustments due to change to works instructions;
- (cl. 9.7) 'an event attributable to the Client or its agents' adding costs and expenses to the works, entitling the contractor to apply for an adjustment (additional payment).

6.7   There are no 'fluctuations' clauses in either version of the contract, therefore the fact that the price of materials or labour may have changed since the contractor tendered would not be reason for the contract price to change.

Contractor's proposed changes

6.8   As part of the Collaborative Working section, the contractor is encouraged to propose changes to the works that will improve quality and/or reduce the contract price (cl. 3.4). The clause then states that if such a proposal is made, the contract administrator will consider the proposal with the client, and may either reject it or issue instructions regarding its implementation. This clause has been rewritten since the previous version, where the onus was on the client to accept or reject it, and which had implied that the matter could be agreed between the contractor and the client before the contract administrator learns of it. The new wording places the authority for the decision firmly in the hands of the contract administrator. However, the contract administrator would be wise to always consult the client as required before any proposal is accepted, particularly if it is likely to have a significant impact on the design or on other aspects of the project.

6.9   Clause 3.5 states that any cost saving is to be divided equally between the parties. The contractor should, therefore, show in its proposal exactly how any savings are calculated. If there is any subsequent negotiation of the savings, the final reduction should be recorded before the instruction is issued.

Costs resulting from instructions regarding tests, etc.

6.10   Under clause 5.5, the contract administrator may instruct that work is uncovered or tested (see paras 5.19–5.22). If the work turns out to be in accordance with the contract, the costs resulting from the instruction are to be added to the contract price. The costs would include not only those of the uncovering and the tests themselves, but also for reinstating the work, making good any damage and disruption to general progress. If these are likely to be significant, the contract administrator could ask the contractor for an assessment of costs before issuing the instruction, in order to fully inform the client regarding the risks and benefits of the instruction prior to proceeding.

6.11   If the contract administrator wishes to accept work that does not accord with the contract, this would have to be done through issuing a change to works instruction. The contract administrator should obtain the agreement of both parties, and ensure that an adjustment to the contract price is agreed and confirmed in writing. The client may not be happy to accept the work (or may later forget that it agreed to it). The contractor may prefer to correct the work, especially if the proposed reduction in the contract price is significant, and in general it cannot be denied the opportunity to do so (*Mul v Hutton Construction Limited*).

Mul v *Hutton Construction Limited* [2014] EWHC 1797 (TCC)

This case concerned what constitutes an 'appropriate deduction' when an employer decided to accept non-conforming work. Although decided in relation to a JCT contract (the JCT Intermediate Building Contract 2005), it is nevertheless applicable to the RIBA Building Contracts. The project concerned an extension and refurbishment work to a country house. A practical completion certificate was issued with a long list of defects attached, and during the defects liability period the employer decided to have this work corrected by other contractors. The employer then started court proceedings against the contractor, to claim back the costs of this work.

A key issue was the interpretation of clause 2.30, which provides that the contract administrator can instruct the contractor not to rectify defects and 'if he does so otherwise instruct, an appropriate deduction shall be made from the Contract Sum in respect of the defects, shrinkages or other faults not made good'. In this case the contractor argued that an 'appropriate deduction' was limited to the relevant amount in the contract rates or priced schedule of works. The court disagreed. It decided that 'appropriate deduction' under clause 2.30 meant 'a deduction which is reasonable in all the circumstances', and could be calculated by any of the following: the contract rates or priced schedule of works; the cost to the contractor of remedying the defect (including the sums to be paid to third-party subcontractors engaged by the contractor); the reasonable cost to the employer of engaging another contractor to remedy the defect; or the particular factual circumstances and/or expert evidence relating to each defect and/or the proposed remedial works.

However, the judge also pointed out that the employer will still have to satisfy the usual principles that apply to a claim for damages, which include showing that it mitigated its loss. If the employer unreasonably refused to let the contractor rectify defects, then it is likely to find its damages limited to what it would have cost the contractor to put them right.

Change to works instructions

6.12   The process for assessing an adjustment to the contract price as a result of a change to works instruction is covered by clauses 5.11–5.13. Clause 5.11 states:

> Subject to clause 5.10, upon receiving a change to Works instruction, including where relevant an instruction under clause 5.9, the Contractor shall calculate the effect (if any) of the instruction on the Contract Price and/or the Date for Completion, and submit details to the Architect/Contract Administrator.

6.13   The contract then states: 'the Contract Administrator and the Contractor aim to agree on any Revision of Time and/or additional payment implications promptly' (cl. 5.12). If the contractor fails to submit the calculation within the 7 days, or the contract administrator and contractor are unable to agree the revision/payment, 'the Architect/Contract Administrator will determine the appropriate adjustment' (cl. 5.13). This requirement is not discretionary, but imperative: the contract administrator must make this determination.

Additional payment

6.14   Clause 9.7 is headed 'Additional Payment' and states:

> If an event attributable to the Client or its agents adds costs to the Works, the Contractor shall inform the Architect/Contract Administrator and shall apply for an adjustment to the Contract Price (with supporting documentation) within a reasonable time of its occurrence.

This wording is far less stringent than the 2014 edition, which required the application for the additional payment to be made within 10 days of the event ending (if a single event) and also stated that if the contractor failed to adhere to these time periods, the right to the additional payment would be lost. Nevertheless, the application must be made 'within a reasonable time', and it is suggested that this should be understood as 'promptly'. It is reasonable that the contract administrator should be told as soon as the event is likely to happen – in that way the contractor and contract administrator can work together to resolve the problem.

6.15   The 'additional payment' clauses are intended to be a means of dealing with what is usually referred to as 'loss and/or expense'. A clause 9.7 'event' is not defined in the RIBA Building Contracts (unlike JCT contracts, there are no listed 'relevant matters'); however, it is suggested that this term is wide enough to include any action by the client or its agents (which would include the contract administrator), including the issue of a change to works instruction. There is, therefore, some possible overlap between clause 9.7 and clauses 5.11–5.13. However, in practice the contractor is likely to apply for all the consequences of a change to works instruction together (as the time limits are the same) and, as long as the contract administrator takes care not to duplicate any award for losses, there should be no difficulty. A similar procedure is followed regarding applications to that for revisions of time, and for change to works instructions, i.e. the contract administrator and contractor endeavour to agree, otherwise the contract administrator determines the appropriate amount (cl. 9.9, see paras 4.30–4.32 and 6.12–6.13 above). If the parties have agreed any rules, then obviously these should be used for the assessment where appropriate. Any failure to give early notification that results in an avoidable increase in the 'costs and expenses' could be taken into account when making the assessment.

Certification and payment

6.16   As noted above, CBC and DBC include several payment options. The default system is the usual one of payment at monthly intervals, but there is also an optional clause (cl. 18) providing for milestone payments or for a single payment on practical completion. CBC contains a further optional clause (cl. 19) providing for advanced payment.

6.17   Whether to select one payment following practical completion rather than monthly payments will depend upon the length of the project. For small works planned to take less than around six weeks, a single payment might be the most convenient for both parties. However, in the case of CBC, it should be noted that for any contract over 45 days this would not comply with the Housing Grants Act. The Contract Details entry at S2 indicates that it should only apply for shorter periods.

6.18   All procedures for payment in CBC are intended to comply with those set out in the Housing Grants, Construction and Regeneration Act 1996 (as amended) (the Housing Grants Act); see Appendix 1.

Advanced payment

6.19   Optional clause 19 in CBC provides for the client to make a payment in advance of the start date. item T of the Contract Details requires the parties to insert the amount, the date for payment and a schedule of repayment instalments, giving dates and amounts. It also provides an option for requiring an advance payment security. This would normally be in the form of a bond, and clause 19.2 states that the client is not liable for any advance payment until 'the Contractor has met the stated requirements for the advanced payment bond'. It is suggested that the advanced payment provisions are only used with a security, as making an advanced payment to the contractor is a significant risk for the client (e.g. the contractor could become insolvent before the monies are expended on the client's behalf). Although many contractors will argue that they are required to pay out large amounts in advance to their subcontractors and suppliers, and are therefore out of pocket, it is generally better that they should take this risk. If a significant reduction in price is offered to the client, the client should consider the risks carefully before making a decision.

Interim payments – monthly certification

6.20   The dates for interim payment where monthly certification is used are as set out in the Contract Details item K (cl. 7.1), which requires the 'First Interim Payment Date' to be entered, and the intervals for subsequent payments. If no first date is inserted, the default is 30 days after the start date, and if no interval is stated, the default is monthly. It is suggested that 'monthly' should be understood as on the same calendar date each month, not four-weekly.

6.21   The 'Interim Payment Date' is defined as 'the date specified in clause 7 and item K of the Contract Details representing the date on which the amount of any payment due under the Contract is calculated and becomes due'. In the CBC the definition also explains that the interim payment date corresponds to the 'Due Date' as described in the Housing Grants Act (which requires all contracts to establish due dates for payment). It should be noted that inserting alternatives to the default periods, i.e. including a longer or shorter period before the first due date or increasing or reducing the intervals, would not constitute a breach of the Act, provided there were still regular payment dates.

6.22   The contractor is entitled to submit an application for payment 'which complies with the requirements of clause 7.5', but must do so at least 7 days before the relevant interim payment date (cl. 7.2). The contractor may prefer to do this, rather than simply wait for a certificate, as it gives it an opportunity to put forward what it thinks the correct figure should be, and why. This information may be very helpful to the contract administrator, but it is not binding; there is no contractual obligation on the contract administrator to accept it, or to provide a detailed rebuttal if it does not agree with the figures put forward (see Figure 6.1). The contract administrator should remember that this document is of no contractual effect, and should always ensure that the amount shown on a payment certificate is an independent and accurate assessment – the figures shown on the contractor's version should never be simply 'rubber stamped'.

6.23   Clause 7.3 states:

> No later than 5 days after each Interim Payment Date, the Architect/Contract Administrator will send the Parties a Payment Certificate which complies with the requirements of clause 7.5

Clause 7.5 then states:

> Any application for payment, Payment Certificate or Payment Notice shall state the:

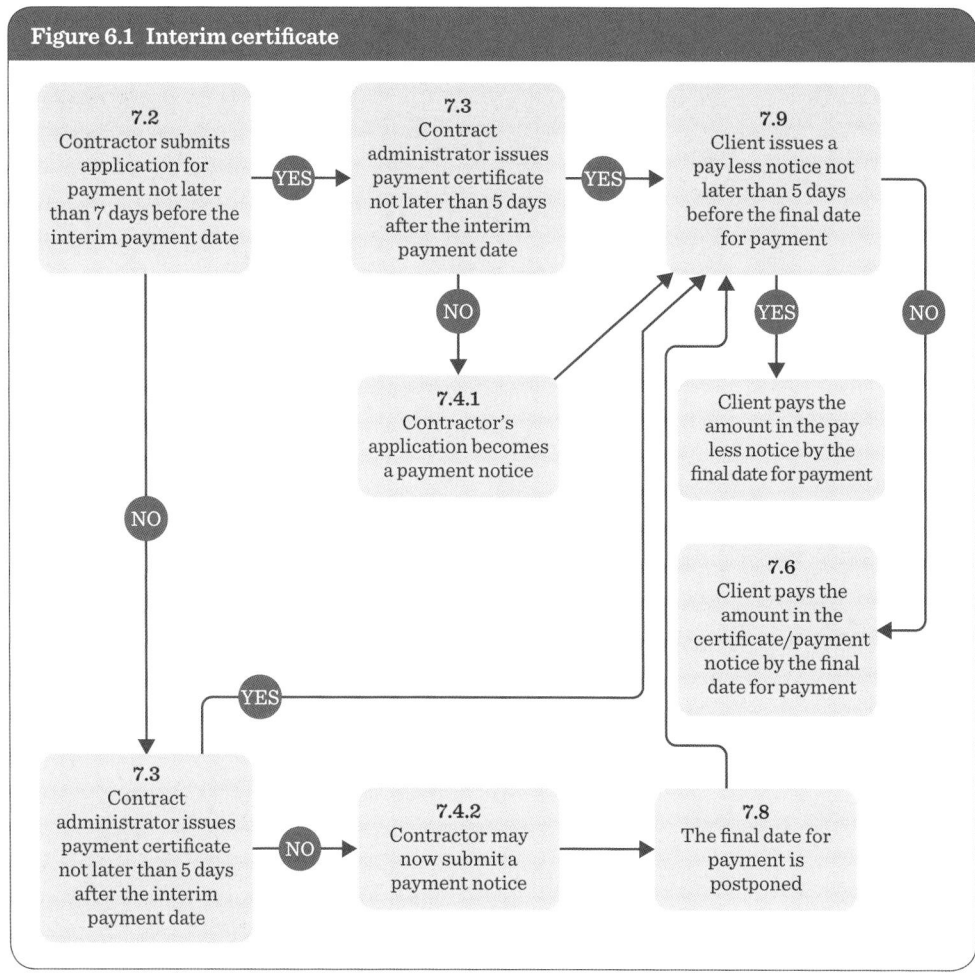

Figure 6.1 Interim certificate

7.5.1 date and period to which it relates

7.5.2 value of work completed in accordance with the Contract

7.5.3 amount of any adjustments required in accordance with the Contract

7.5.4 amount of any Retention

7.5.5 total amount included in previous Payment Certificates or Payment Notices

7.5.6 balance due (from the Client to the Contractor or from the Contractor to the Client) at the relevant payment due date.

The contract administrator should note that it does not have the power to include or deduct amounts not set out in the clause, therefore the calculation set out above should be followed precisely when preparing the certificate. Some particular matters are discussed below.

The total value of work carried out

6.24 Note that this should only be work that is in accordance with the contract, and the contract deliberately does not mention materials and goods that are stored on or off site but not yet incorporated in the works, nor does it mention prefabricated off-site items, therefore neither of these should be included in the certificate. This is an important difference to many other standard contracts, which typically require that at least on-site materials etc. are included in the payment certificate.

6.25 The RIBA Building Contracts, unlike some others, do not include 'property vesting' clauses, which provide that such materials and goods will become the property of the client once certified, even if, for example, the contractor has not yet paid its suppliers. Once materials have been fixed to the construction, they will become the property of the client, and cannot be removed by the contractor or a subcontractor. The client could be at risk, however, where materials have not yet been built in, even where the materials have been certified and paid for. The contractor might not actually own the materials paid for because of a retention of title clause in the contract for the sale of the materials. Under the Sale of Goods Act 1979, sections 16 to 19, property in goods normally passes when the purchaser has possession of them, but a retention of title clause will be effective between a supplier and a contractor even where the contractor has been paid for the goods, provided they have not yet been built in.

6.26 Without property vesting clauses there is, therefore, a danger that unpaid suppliers may return and remove unfixed goods, despite the fact that the client has paid the contractor for them. As there is no requirement in the RIBA Building Contracts to include the value of any of these in interim certificates, normally it would be inadvisable to do so.

6.27 The contract administrator should only certify after having carried out an inspection to a reasonably diligent standard, and should not include any work that appears not to have been properly executed, whether or not it is about to be remedied, and even if the retention is adequate to cover any anticipated remedial work; retention is to cover latent (i.e. hidden) defects, not patent defects (those that are apparent, see *Townsend* v *Stone Toms* and *Sutcliffe* v *Chippendale & Edmondson*). Contract administrators should also note the case of *Dhamija* v *Sunningdale Joineries Ltd*, which stated that a quantity surveyor is not responsible for determining the quality of work executed.

> *Townsend* v *Stone Toms & Partners* (1984) 27 BLR 26 (CA)
>
> Mr Townsend engaged Stone Toms as architect in connection with the renovation of a farmhouse in Somerset. John Laing Construction Ltd was employed to carry out the work on JCT67 Fixed Fee Form of Prime Cost Contract. Following the end of the defects liability period, the architect issued an interim certificate that included the value of work which had already been included in the schedule of defects, and which the architect knew had not yet been put right. Mr Townsend brought proceedings against both Stone Toms and Laing. Laing made a payment into court of £30,000, which was accepted by Mr Townsend in full and final settlement. Mr Townsend then continued with the proceedings against the architect, claiming that he was entitled to recover any excess that he might have obtained from Laing had he continued with those proceedings. The Official Referee assessed the total value of the claims against Laing as only £25,000, therefore no excess was recoverable. The Deputy Official Referee also found that the architect was not negligent in issuing the interim certificate. Mr Townsend appealed and the Court of Appeal, although approving the lower court's decision on the effect of the payment into court, held

that the architect had been negligent. Oliver LJ stated (at page 46): the whole purpose of the certification is to protect the client from paying to the builder more than the proper value of the work done, less proper retention, before it is due. If the architect deliberately over-certifies work which he knows has not been done properly, this seems to be a clear breach of his contractual duty, and whether certification is described as 'negligent' or 'deliberate' is immaterial.

Sutcliffe v Chippendale & Edmondson (1971) 18 BLR 149

(Note: this case is the first instance decision which was appealed to the Court of Appeal sub nom. *Sutcliffe* v *Thackrah*)

Mr Sutcliffe engaged the architect Chippendale & Edmondson in relation to a project to build a new house. No terms of engagement were agreed, but the architect proceeded to design the house, invite tenders and arrange for the appointment of a contractor on JCT63. Work progressed slowly and towards the end of the work it became obvious that much of the work was defective.

The architect had issued ten interim certificates before Mr Sutcliffe entirely lost confidence, dismissed the architect and threw the contractor off the site. He then had the work completed by another contractor and other consultants at a cost of around £7,000, in addition to which he was obliged, as a result of the original contractor having obtained judgment against him, to pay all ten certificates in full. As this contractor was then declared bankrupt, Mr Sutcliffe brought a claim against the architect. The architect contended, among other things, that its duty of supervision did not extend to informing the quantity surveyor of defective work that should be excluded from the valuation. His Honour Judge Stabb QC found for Mr Sutcliffe, stating 'I do not accept that the words "work properly executed" can include work not then properly executed but which it is expected, however confidently, the Contractor will remedy in due course' (at page 166).

Dhamija v Sunningdale Joineries Ltd, Lewandowski Willcox Ltd, McBains Cooper Consulting Ltd [2010] EWHC 2396 (TCC)

An action was brought against the building contractor, the architect and the quantity surveyor (QS) (McBains) for defects in the design and construction of a home. There had been no written or oral contract with the QS, so the terms of its engagement would be those that would be implied. It was held that a QS's contract of retainer would include an implied term that the QS acts with the reasonable skill and care of a QS of ordinary competence and experience when valuing the works properly executed for the purposes of interim certificates, but that the QS would not owe an implied duty to exclude the value of defective works from valuations, however obvious the defects. This was the exclusive responsibility of the architect appointed under the contract. Further, the QS owed no implied duty to report the existence of defects to the architect.

6.28   Where work that has been included in a payment certificate subsequently proves to be defective, the value can be omitted from the next certificate. Under clause 7.5.6 the contract administrator has the power to issue a 'negative' certificate, should this be necessary to correct an earlier over-valuation.

6.29   It should be noted that clause 7.5.1 does not specifically state at what point in time the value of work should be assessed. The definition of 'Interim (or Final) Payment Date' as 'the date on which the amount of any payment due under the Contract is calculated' suggests that the valuation should assess the work properly carried out at the interim payment date. This could create a slight problem in practical terms. Normal practice is that the contract administrator makes its valuation immediately after an inspection (in order to ensure that only work correctly carried out is included), and will then issue the certificate a few days later. If the contract administrator wishes to issue its certificate on the interim payment date (in order to allow the client the maximum payment period), this will mean that the inspection will be a few days before the interim payment date, and therefore will not reflect the value at that date. However, provided the gap is kept to a minimum, the value shown on the certificate will be close to the value at the interim payment date. The alternative, i.e. for the contract administrator to attempt to guess exactly what work will have been carried out between the inspection/valuation and the payment date, would place the client at risk.

Any adjustments required in accordance with the contract

6.30   Clause 7.5.3 also requires the contract administrator to certify 'amount of any adjustments required in accordance with the Contract'. This would cover amounts due to the contractor that are not related to the value of the works, for example the costs referred to in clause 9.7 (see para. 6.14).

6.31   There are three further areas of adjustment that should be noted:

● Failure to make good defective work – under clauses 5.8 and 10.6, the client may take action if the contractor fails to make good defective work and the contractor 'shall be responsible for all costs incurred' (cl. 5.8.4 and 10.6).

● Failure to take out insurance – under clause 6.4, the 'Contract Price will be adjusted' if either party fails to provide evidence that it has taken out the required insurance, and the other takes it out as required.

● Liquidated damages – clause 10.1 states that if the contractor does not complete by the relevant date for completion, the client 'shall be entitled to deduct Liquidated Damages'.

6.32   In the case of the first deduction, the contract does not state how the costs are to be recovered. Assuming the work is, as a result of the client's action, now complete to the standards in the contract, the certificate could include this work (but making the circumstances clear), taking account of the net effect on the contract price (normally a deduction, i.e. a net additional cost to the client). However, it may be more practical for the contract administrator to simply omit the work, and for the client to deduct the net additional costs from the certified amount. For the second item, however, the contract specifically states that this requires an adjustment to the contract price, therefore this cost should be accounted for in the Certificate.

6.33   With the third, although there does not appear to be any reason why the deductions cannot be made on the face of the payment certificate, in practice it is more usual, and more sensible, for the client to make a deduction by means of a pay less notice (see para. 6.50). Whether to make the deduction is entirely a decision for the client, so if it is made on the certificate, the contract administrator must consult with the client and ensure that there is a written record of the client's decision. Alternatively, for clarity the contract administrator could note on the certificate what client deductions have been made from earlier certified amounts; however, this would simply be a record for information, and would not affect the calculation of the total certified amount.

6.34   Note that if optional clause 19 has been selected (CBC only), which allows for the client to make a payment in advance (see para. 6.19), the parties will have set out a schedule of repayment instalments in item T of the Contract Details. Clause 19.1 states that the client should deduct the repayments *from* payment certificates, so there is no need for the deduction to be included *within* the certificate. As with other deductions, the client should follow the pay less notice procedures as discussed below.

Amount of any retention

6.35   The definition of terms defines 'Retention' as 'a percentage of the amount included in a Payment Certificate that is deducted from a payment in accordance with clause 7'. Under clause 7.14 the retention is set at 5 per cent up until practical completion, and 2.5 per cent during the defects fixing period. The wording does not make it clear exactly what the percentage is deducted from, but normal practice would be to deduct it from the value of work properly carried out, before the other deductions are made (e.g. as under JCT contracts). There is no opportunity to specify a different rate to 5 per cent; if a different rate is preferred then the contract would need to be amended.

The total amount included in previous payment certificates or payment notices

6.36   The application or payment certificate should state 'total amount included in previous Payment Certificates or Payment Notices' (cl. 7.5.5). This might include payments made to the contractor as a result of a payment notice, in situations where the contract administrator has failed to issue a payment certificate (see para. 6.54). Care should be taken when calculating the final amount due; for example, if the total deduction is made before retention is deducted, it should be the total gross value of work identified on the previous payment certificate, and before any other deductions are made.

6.37   Note that there should not be any other payments to the contractor which are not covered by a payment certificate or payment notice. Sometimes, on smaller projects, the client may ask the contractor to, for example, purchase items of equipment not included in the contract documents. This type of direct arrangement should be avoided; all items should be handled through the contract administrator by means of a change to works instruction, as otherwise there are issues as to whether this forms part of the works and is therefore covered by such matters as liability for defects and insurance. Should such situations inadvertently arise, clause 7.5 does not entitle the contract administrator to deal with them under the main contract.

Interim payments – milestone payments

6.38   As an alternative to the monthly due dates, the contract allows for payment by milestones (cl. 18). If adopted, the parties must define the milestones in detail (item S1 in the Contract Details). The milestones will normally be identifiable points in the completion of the project, for example completion of: (1) foundations and groundworks, (2) ground floor slab and walls up to the damp-proof course, (3) external walls, (4) roof, etc.

6.39   Linking payment to milestones introduces an incentive to the contractor to maintain steady progress throughout the project (whereas the threat of liquidated damages applies only to achieving practical completion).

6.40   The exact work that must have been correctly completed for a milestone to be achieved should be set out in detail in item S1. In addition, the parties are required to set out the payment that will be made on achievement of the milestone, either as a value or as a percentage of the contract price.

6.41   The corresponding optional clauses for item S1 state:

> 18.1 If item S1 or item S2 of the Contract Details is selected, then the part of item K of the Contract Details regarding payment certificate frequency shall not apply.
>
> 18.2 If item S1 of the Contract Details is selected, then the Interim Payment Date shall be the dates on which the milestones are achieved.
>
> 18.3 When the Architect/Contract Administrator is satisfied that a milestone has been achieved it will notify the Parties.

6.42   Clause 18 will need to be applied with care. The milestone payment certificates will state the value or percentage for that stage, once the milestone has been reached, and should comply in all respects with clause 7.5, including accounting for any adjustments required under the contract. For the milestone to be achieved, the work should not simply be complete, but be complete to the standard set out in the contract, with no patent defects. Item S1 in the Contract Details requires the parties to insert a 'date to be achieved by' for each milestone. However these operate as a target, no amounts would be certified until the milestone is actually achieved.

Interim payments – payment on practical completion

6.43   Optional clause 18.4 states:

> If item S2 of the Contract Details is selected, the Interim Payment Date shall be 7 days after the Architect/Contract Administrator has certified Practical Completion.

6.44   Although the clause does not specifically state this, it is clear from the guidance notes that this is intended to replace monthly certification with a single payment at practical completion. As such, it should only be selected when the contract period is very short, and no longer than 45 days (as otherwise the contract must comply with the Housing Grants Act). As this is an 'interim payment' the contract administrator must issue a payment certificate in accordance with clause 7.3 as discussed above. This payment certificate would, therefore, show the adjusted contract price, with any applicable adjustment as listed in clause 7.5, and with 2.5 per cent retention.

Payment of interim payments

6.45   The client (or contractor, if the certificate shows a balance due to the client) is required to pay the amount shown on a payment certificate 'by the Final Date for Payment' (cl. 7.6). 'The Final Date for Payment' is defined as 'the date, specified in clause 7, following the due date for interim or final payment, by which a payment that is due should be paid'.

Clause 7.7 states that 'the Final Date for Payment of an interim and the final payment shall be 14 days after the relevant payment due date'. The problem with this wording, apart from being somewhat circular, is that 'the relevant payment due date' is not a defined term, or otherwise set out in the contract. However, bearing in mind that the 'Interim (or Final) Payment Date' is defined, at least in CBC, with reference to the Housing Grants Act requirement for 'due dates', it is likely that the Final Date for Payment of an interim certificate will be understood as 14 days after the interim payment date.

6.46   The time periods are quite short: if the contract administrator takes the full 5 days allowed after the interim payment date to issue the certificate, then this will leave the client with only 9 days to pay, which could of course include 4 weekend days; clearly, the contract administrator should issue payment certificates promptly. If the certificate shows a balance due from the contractor to the client, then the contractor must similarly pay the amount within 14 days.

Contractor's invoice

6.47   Clause 7.15 requires the contractor to issue the client with a valid VAT invoice, and the client to pay the invoice promptly. The rules relating to VAT are beyond the scope of this Guide; if advice is needed the parties should contact HM Revenue & Customs or an appropriate expert. VAT is a matter of law, and any mistakes or attempt to avoid it would be a breach of the law; in this case it would also be a breach of contract, allowing the parties to claim against each other for losses due to an infringement.

6.48   It is suggested that the requirement to pay the VAT invoice 'promptly' does not override the time limit as set out in clause 7.6; the client should pay the amount shown on the payment certificate within 14 days of the interim payment date, regardless of when the VAT invoice is issued. If it is issued shortly after the certificate, the VAT invoice could be paid at the same time; if it arrives later, the VAT will be paid separately.

Pay less notices

6.49   If the client or contractor wishes to pay less than the certified amount, it must issue a 'Pay Less Notice' not later than five days before the final date for payment (cl. 7.9). The pay less notice should state the amount the client considers due, and how it was calculated. If the certificate shows a balance due to the client, then the contractor may issue a pay less notice (cl. 7.9). The contract does not expressly state so, but it can be assumed the paying party can pay the lesser amount shown in the notice, provided it was issued in accordance with the contract terms.

6.50   The client has the right to pay less than the certified amount, but only for reasons that can be justified under the terms of the contract. It is suggested that these could include that some of the work covered by the certificate was not in accordance with the contract, or if there was some error in the calculation of the payment certificate, or for any of the matters which might entitle the client to a deduction, as noted at para. 6.31, namely:

- failure to make good defective work;
- failure to take out insurance;
- liquidated damages.

6.51   In the case of the contractor, a pay less notice will only arise if the certificate shows a negative balance. This is only likely to happen if work which has previously been certified later proves defective and is excluded from a subsequent certificate. It is possible that the contractor may disagree with this decision and refuse to repay the required amount. If this happens, the client's remedy would be to wait until the next certificate, by which time the balance may be due to the contractor, or to take the matter to adjudication

Contractor's remedy if no certificate issued

6.52   The contractor is given various remedies should the above payment systems break down. If no payment certificate is issued, and the contractor had made an application for payment under clause 7.2, then this is said to become a 'Payment Notice' (cl. 7.4.1). The contractor may need to forward the application to the client, if it has not already done so. The client must pay this amount on the final date for payment for interim certificates (cl. 7.6), subject to any pay less notice.

6.53   If no application had been made, the contractor may now issue a payment notice, showing the amount it considers due and how it was calculated (cl. 7.4.2). In this case, the final date for payment is postponed by the same number of days that it took the contractor to issue the payment notice after the final date that the certificate should have been issued (cl. 7.8).

6.54   In either case the client should notify the contract administrator that payment has been made, so that this can be taken into account in the next certificate. A similar system is set up for the final payment certificate; if the contract administrator fails to issue the certificate, the contractor may submit a final payment notice (cl. 7.12.2 and 7.13).

Contractor's remedies if certificate not paid

6.55   If the client fails to pay an amount that is due by the final date for payment, clause 7.16 makes provision for simple interest to accrue on any unpaid amount. The rate of interest can be set by the parties (e.g. they could adopt the rate stipulated in many JCT contracts, i.e. five per cent over the base rate of the Bank of England). If no rate is set in item K, the default is the statutory right to interest under the Late Payment of Commercial Debts (Interest) Act 1998. The interest accrues from the final date for payment until the amount is paid. The provisions apply to the final payment certificate as well as to interim certificates.

6.56   If the client makes a valid deduction following a pay less notice, it is suggested that interest would not be due on this amount. The clause does not refer to the amount stated on the certificate but to 'any unpaid amount', which would take into account valid deductions.

Right of suspension

6.57   The contractor is given a 'right to suspend' under clause 8.1. If the client fails to pay the contractor by the required time limit, the contractor has a right to suspend performance of some or all of its obligations under the contract, which would not only include the carrying out of the work, but could also, for example, extend to any insurance obligations. In addition, under optional clause 22, the contractor may request that the client provides evidence of its ability to pay the contract price, and may suspend its obligations under clause 8 if the evidence is not provided.

6.58   This non-payment is stated to be of 'an amount that is due', therefore the contractor may not suspend work if a valid pay less notice has been issued by the client. The contractor must have given the client a written 7-day notice of its intention to suspend work and stated the grounds for the suspension, and the default must have continued for a further 7 days. The contractor must resume work when the payment is made.

6.59   Under these circumstances the suspension would not give the client the right to terminate the contractor's employment. The suspension will, however, give the contractor the right to a revision of time, and to reasonable expenses and costs arising out of the suspension (cl. 8.2). This right is required by the Housing Grants Act, and therefore could be deleted in DBC if the client would prefer.

Termination

6.60   The contractor also has the right to terminate its employment if the client does not pay amounts due (cl. 12.3), at least to the extent that the non-payment might be considered a material breach. The contractor must have given a 14-day notice of this intention, which must specify the default and refer to the specific clause. It should be noted that this right is probably limited to non-payment of significant amounts, or to persistent non-payment, and so could not be exercised for minor non-payment. It is unlikely, in practice, that the contractor would terminate for minor shortfalls, given that the suspension remedy is available.

7

Practical completion, completion and post-completion

Practical completion

7.1   The decision to certify practical completion is one of the most important that the contract administrator makes during the whole project as it triggers many contractual consequences that are important to both the client and the contractor. The period leading up to practical completion can be difficult and stressful. Sometimes this may be due to an (erroneous) belief by the contractor that the word 'practical' indicates that something that is 90 per cent finished, or could be occupied, has reached practical completion. At other times there may be pressure from a client who is very anxious to occupy the building, and who may not appreciate how much work is still needed to correct what might appear to minor matters. However, in the RIBA Building Contracts (unlike many other contracts), a clear definition of practical completion is given. Clause 9.10 states:

> For Practical Completion to occur, the following must apply:
>
> 9.10.1 any requirements stated in the Contract Documents and required by law shall have been satisfied
>
> 9.10.2 no aspect of the Works or Section of the Works shall be outstanding
>
> 9.10.3 the Works shall be uncluttered and safe.

7.2   In addition, if optional clause 23 of CBC is selected, the contractor must have provided copies of all collateral warranties/third-party rights agreements required before practical completion can be certified (cl. 23.2).

7.3   It is suggested that 'no aspect of the Works or Section of the Works shall be outstanding' includes aspects of quality, as well as quantity, therefore if the quality of work is unsatisfactory, practical completion as defined has not been reached. This interpretation has been supported by the courts, for example in the well-known case of *H W Nevill (Sunblest) Ltd* v *William Press & Son Ltd*. If the client nevertheless wishes to occupy the building with minor work outstanding, then a special arrangement will need to be made, as discussed below (see para. 7.9).

> *H W Nevill (Sunblest) Ltd* v *William Press & Son Ltd* (1981) 20 BLR 78
>
> Here William Press entered into a contract with Sunblest to carry out foundations, groundworks and drainage for a new bakery on a JCT63 contract. A practical completion certificate was issued, and new contractors commenced a separate contract to construct the bakery. A certificate of

making good defects and a final certificate were then issued for the first contract, following which it was discovered that the drains and the hardstanding were defective. William Press returned to the site and remedied the defects, but the second contract was delayed by four weeks and Sunblest suffered losses as a result. It commenced proceedings, claiming that William Press was in breach of contract and in its defence William Press argued that the plaintiff was precluded from bringing the claim by the conclusive effect of the final certificate. Judge Newey decided that the final certificate did not act as a bar to claims for consequential loss. In reaching this decision he considered the meaning and effect of the certificate of practical completion and stated (at page 87): 'I think that the word "practically" in clause 15(1) gave the architect a discretion to certify that William Press had fulfilled its obligation under clause 21(1) where very minor *de-minimis* work had not been carried out, but that if there were any patent defects in what William Press had done then the architect could not have issued a certificate of practical completion.'

7.4   It is open to the parties to set out their own particular requirements for practical completion, including the standard expected and the means for establishing if it has been reached (cl. 9.10.1). The parties might consider, for example, requiring that mechanical services are properly commissioned, that any performance in use criteria are tested and checked (e.g. airtightness and acoustic requirements), that an operations manual is handed over and the client trained in operation of the building. Any of these would, of course, have to have been made clear in the tender documents.

7.5   The contractor is required to notify the contract administrator when it thinks that practical completion of the works, or a section, has been achieved (cl. 9.11). If the contract administrator agrees, it will issue a certificate of practical completion of the works or section as appropriate (cl. 9.11.1).

7.6   If the contract administrator does not agree, unlike in the 2014 edition, there is no requirement to inform the contractor of this. Although it may nevertheless be sensible to let the contractor know that one or more of the conditions in clause 9.10 has not yet been met, it is important to note that the contract administrator is not required to give reasons for this decision, and there is certainly no obligation to produce a 'snagging list'. It is common practice for the contract administrator to issue such a list but, unless it has entered into special terms of appointment with the client, there is no need for it to do so. Not only is it very time-consuming and resource-hungry, it is effectively taking on the contractor's quality assurance duties, which could ultimately lead to a confusion as to roles and responsibilities. If the contract administrator is concerned about particular defects, there is no harm in informing the contractor, provided it is made clear that these are just examples of some of the shortfalls on the project and that they are not intended to be a comprehensive list.

Consequences of practical completion

7.7   The consequences of practical completion are as follows:

- half the withheld retention is released (cl. 9.11.2);
- liquidated damages will cease (cl. 10.1);
- the defects fixing period commences (cl. 10.2);
- the contractor remedies defects notified during the defects fixing period (cl. 10.3).

7.8   These consequences are important and therefore the contract administrator should take great care to ensure the defined level of completeness has been reached before certifying practical completion. Issuing the certificate with work outstanding places considerable additional risk on the client: the key areas being that there is no longer the sanction of liquidated damages to encourage the contractor to finish promptly, and the client holds only half the retention sum as security against any hidden defects. In addition, matters such as insurance of the works and health and safety will need to be resolved, and there are practical issues to do with managing the programming and payment for the outstanding work. The contracts have no provisions to cover these aspects as they assume that the work will be finished, with the exception of defects that appear during the defects fixing period.

Use/occupation before practical completion

7.9   If the works have not reached practical completion, but the client wishes to use them, the contracts contain a provision that may be of help to the client. This allows the client to request to 'take over' any part or parts of the works or a section of the works before the contract administrator certifies practical completion (cl. 9.12). The contractor must grant permission, but only if the use does not interfere with the carrying out of the works. Practical completion is not certified, but the contract administrator must issue a notice 'clearly identifying the part(s) taken over and the date of takeover, as well as any outstanding work and arrangements for access by the Contractor' (cl. 9.13.1).

7.10   In addition, the following consequences apply:

> 9.13.2 Practical Completion will be deemed to have taken place on the date of takeover for the relevant part(s)
>
> 9.13.3 the Defects Fixing Period for that part(s) shall start from that date
>
> 9.13.4 in respect of the Contract Price for that part of the Works, the Contractor is entitled to half the amount currently retained in accordance with clause 7.15.

The contract does not specifically state what will happen to the liquidated damages payable, if the completion date is not met for the remainder of the works. In view of clause 9.13.2, the likely intention was that these would be reduced, and the normal method is to do this in proportion to the value of the work taken into over. However it would be sensible to agree this prior to the part being occupied.

7.11   There can be situations where the client is anxious to occupy part or possibly the whole of the works before any parts are sufficiently complete to be taken over under clause 9.12. Rather than viewing the work as having reached practical completion when it has not, it would be better if the parties make an ad hoc agreement as to what the arrangement will be. A suggestion was put forward in the 'Practice' section of the *RIBA Journal* (February 1992), which has frequently proved useful in practice: in return for being allowed to occupy the premises, the client agrees not to claim liquidated damages during the period of occupation. Practical completion obviously cannot be certified, and the defects fixing period will not commence, nor will there be any release of retention money, until the work is complete. Health and safety will need to be given careful consideration, and matters of insuring the works will need to be settled with the insurers.

7.12   Because such an arrangement would be outside the terms of the contract, it should be covered by a properly drafted agreement that is signed by both parties. (The cases of *Skanska* v *Anglo-Amsterdam Corporation* and *Impresa Castelli SpA* v *Cola Holdings Ltd* illustrate the importance of drafting a clear agreement.) It may also be sensible to agree that, in the event that the contractor still fails to achieve practical completion by the end of an agreed period, liquidated damages would begin to run again, possibly at a reduced rate. In most circumstances this arrangement would be of benefit to both parties, and is certainly preferable to issuing a heavily qualified takeover notice, or a certificate of practical completion, listing numerous incomplete items of work.

Skanska Construction (Regions) Ltd v *Anglo-Amsterdam Corporation Ltd* (2002) 84 Con LR 100

Anglo-Amsterdam Corporation (AA) engaged Skanksa Construction (Skanska) to construct a purpose-built office facility under a JCT81 With Contractor's Design form of contract. Clause 16 had been amended to state that practical completion would not be certified unless the certifier was satisfied that any unfinished works were 'very minimal and of a minor nature and not fundamental to the beneficial occupation of the building'. Clause 17 of the form stated that practical completion would be deemed to have occurred on the date that the employer took possession of 'any part or parts of the Works'. AA wrote to Skanska confirming that the proposed tenant for the building would commence fitting out works on the completion date. However, the air-conditioning system was not functioning and Skanska had failed to produce operating and maintenance manuals. Following this date the tenant took over responsibility for security and insurance, and Skanska was allowed access to complete outstanding work. AA alleged that Skanska was late in the completion of the works and applied liquidated damages at the rate of £20,000 per week for a period of approximately nine weeks. Skanska argued that the building had achieved practical completion on time or that, alternatively, partial possession of the works had taken place and that, consequently, its liability to pay liquidated damages had ceased under clause 17.

The case went to arbitration and Skanska appealed. The court was unhappy with the decision and found that clause 17.1 could also operate when possession had been taken of all parts of the works and was not limited to possession of only part or some parts of the works. Accordingly, it found that partial possession of the entirety of the works had, in fact, been taken some two months earlier than the date of practical completion, when AA agreed to the tenant commencing fit-out works. Consequently, even though significant works remained outstanding, Skanska was entitled to repayment of the liquidated damages that had already been deducted by AA.

Impresa Castelli SpA v *Cola Holdings Ltd* (2002) CLJ 45

Impresa agreed to build a large four-star hotel for Cola Holdings Ltd (Cola), using the JCT Standard Form of Building Contract With Contractor's Design, 1981 edition. The contract provided that the works would be completed within 19 months from the date of possession. As the work progressed, it became clear that the completion date of February 1999 was not going to be met, and the parties agreed a new date for completion in May 1999 (with the bedrooms being made available to Cola in March) and a new liquidated damages provision of £10,000 per day, as opposed to the original rate of £5000. Once the agreement was in place, further difficulties with progress were encountered, which meant that the May 1999 completion date was also unachievable. The parties entered into a second variation agreement, which recorded that access for Cola would be allowed to parts of the hotel to enable it to be fully operational by September

1999, despite certain works not being complete (including the air conditioning). In September 1999, parts of the hotel were handed over, but Cola claimed that such parts were not properly completed. A third variation agreement was put in place with a new date for practical completion and for the imposition of liquidated damages. Disputes arose and, among other matters, Cola claimed for an entitlement for liquidated damages. Impresa argued that it had achieved partial possession of the greater part of the works, therefore a reduced rate of liquidated damages per day was due. The court found that, although each variation agreement could have used the words 'partial possession', they had in fact instead used the word 'access'. The court had to consider whether partial possession had occurred under clause 17.1 of the contract, which provides for deemed practical completion when partial possession is taken, or whether Cola's presence was merely 'use or occupation' under clause 23.3.2 of the contract. The court could find nothing in the variation agreements to suggest that partial possession had occurred. It therefore ruled that what had occurred related to use and occupation, as referred to in clause 23.3.2 of the contract, and that the agreed liquidated damages provision was therefore enforceable.

Non-completion

7.13   If the contractor fails to achieve practical completion of the works by the date for completion, the client may deduct liquidated damages at the agreed rate (cl. 10.1). There is no need for the contract administrator to have issued any non-completion certificate, although it would normally write to the client advising it of the position.

The defects fixing period

7.14   The contractor is required to 'remedy all defects associated with the Works notified to them during the Defects Fixing Period' (cl. 10.3). The period begins at practical completion and lasts for the length of time entered in the Contract Details (cl. 10.2, item M). The notes to item M indicate that the minimum period is 3 months, however, 12 months (12 months is commonly used so that any mechanical services such as heating systems can be run through four seasons). A shorter period might be acceptable for very small projects.

7.15   It is suggested that although clause 10.3 suggests that the contractor is only obliged to correct notified defects, the obligation is wider than this. As part of its general duty to complete the works satisfactorily, the contractor would be obliged to ascertain whether any defects have appeared, and to correct any work that was defective. The notification process allows for the architect to raise matters of concern, and for arrangements to be made for access that suit all parties, but if the contractor is aware of any matters not notified it should alert the contract administrator so that they can be dealt with. There is no express requirement for the contract administrator to prepare a schedule of defects at the end of the defects fixing period, as there is in some of the other standard contracts (e.g. IC16, cl. 2.30). However, the contract administrator is required to issue a notice to the contractor requiring it to remedy any defect *which it fails to fix* (cl. 10.5). If the contractor does not comply with the notice promptly, the client may engage others to rectify the problem, and all costs are the responsibility of the contractor (cl. 10.6). This right would only arise if the correct procedures are followed; if the client, for example, refuses to allow the contractor reasonable access for inspecting and undertaking work, this might result in the client being unable to claim the costs of remedying the defects by others (*Pearce and High Ltd* v *John P Baxter and Mrs A S Baxter*).

Pearce and High Ltd v *John P Baxter and Mrs A S Baxter* [1999] BLR 101 (CA)

The Baxters employed Pearce and High on MW80 to carry out certain works at their home in Farringdon. Following practical completion, the architect issued interim certificate no. 5, which the employer did not pay. The contractor commenced proceedings in Oxford County Court, claiming payment of that certificate and additional sums. The employer in its defence and counterclaim relied on various defects in the work that had been carried out. Although the defects liability period had by that time expired, neither the architect nor the employer had notified the contractor of the defects. The Recorder held that clause 2.5 was a condition precedent to the recovery of damages by the employer, and further stated that it was a condition precedent that the building owner had notified the contractor of patent defects within the defects liability period. The employer appealed and the appeal was allowed. Lord Justice Evans stated that there were no clear express provisions within the contract which prevented the employer bringing a claim for defective work, regardless of whether notification had been given. He went on to state, however, that the contractor would not be liable for the full cost to the employer of remedying the defects, if the contractor had been effectively denied the right to return and remedy the defects itself.

7.16   In practice, correction of defects is normally left until the end of the defects fixing period, as it is usually more convenient for both parties if all the work is done together. However, sometimes a defect can cause considerable problems to the client, in which case the contractor should take steps as soon as it is aware of the matter. In either case, if the contractor fails to deal with any defect satisfactorily, the contract administrator should issue a clause 10.5 notice, in order to trigger the client's right to engage others if necessary. After the end of the defects fixing period, when the contract administrator is satisfied that all defects are remedied, the contract administrator is required to notify the parties accordingly (cl. 10.4).

Payments following practical completion

7.17   Following practical completion, in the absence of any limiting provision it appears as if the payment dates, and therefore certification, continue at monthly intervals (cl. 7.1). With respect to milestone payments, or the single payment on practical completion, as clause 18.1 states that 'the part of item K of the Contract Details regarding payment certificate frequency shall not apply', there would be no further certificates until the final payment, unless the parties agree otherwise.

Final contract price and payment

7.18   Most contracts contain provisions for dealing with the final assessment of the contract price. Clause 7.11.1 requires the contractor to submit its calculation, along with the relevant supporting documentation, to the contract administrator within 90 days of practical completion (note that in the 2014 version the equivalent clause stated within 90 days of the defects fixing period, which must have been an error).

7.19   The contract then requires the contract administrator and the contractor to endeavour to reach agreement on the final amount within 30 days of the submission (cl. 7.11.2). If they are unable to reach agreement, or the contractor fails to submit a calculation, the contract administrator is required to calculate the 'final Contract Price' and notify it to the contractor not later than '14 days after the 30 day period referred to in clause 7.11.2' (cl. 7.11.3).

7.20   The final contract price is said to be 'subject to resolution of any defects arising during the Defects Fixing Period' (cl. 7.11.4). The contract does not explain what this means but, considered in practical terms, at the time the notification is given the full extent of defects may not be known. The time limits are well within the defects fixing period – in fact, there is nothing to prevent the contractor submitting its assessment immediately after practical completion, in which case either agreement will be reached within 30 days or a notification must be sent within a further 14 days – just six weeks into a defects fixing period that could last six months or a year – at a point when work that appears to be sound may

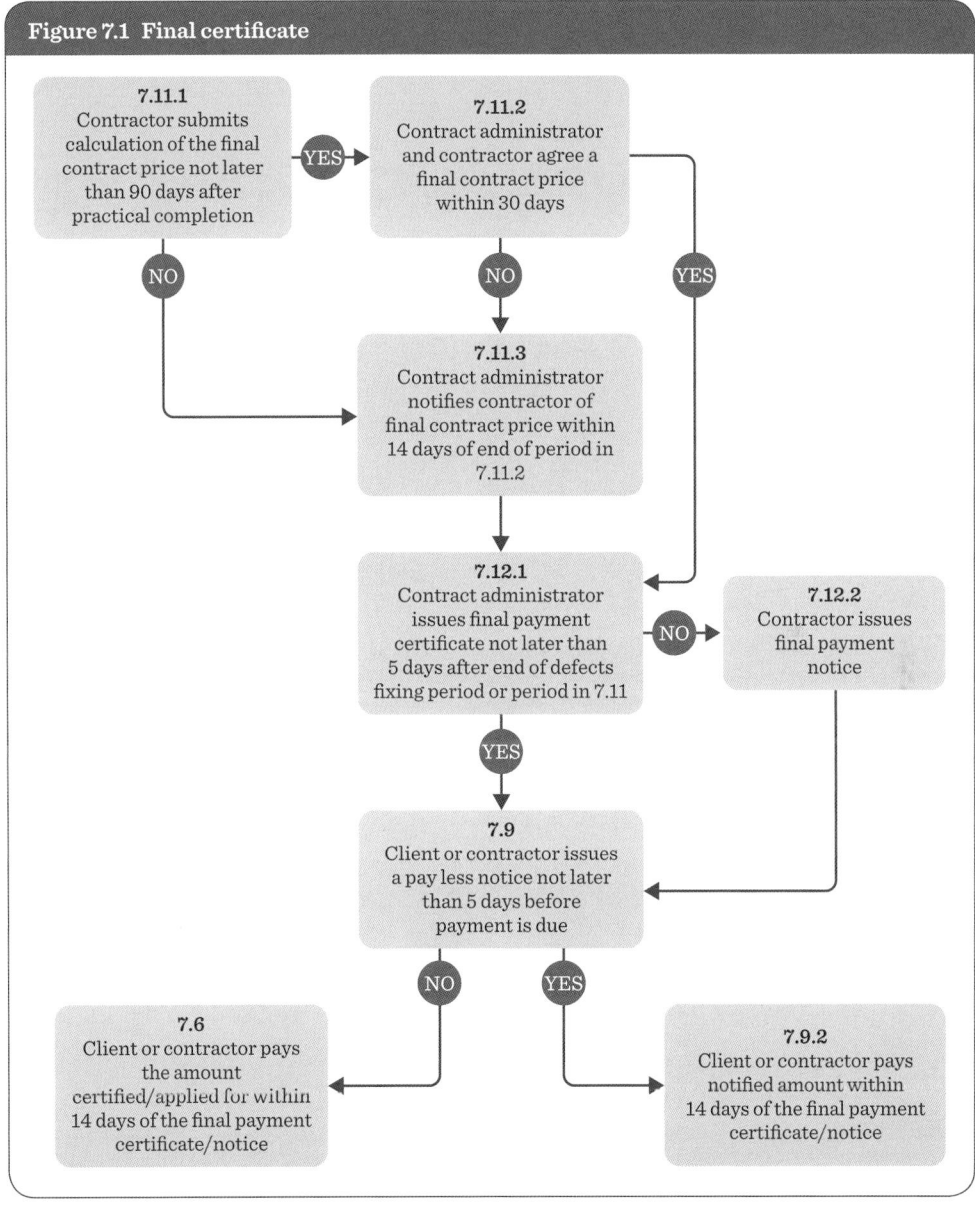

Figure 7.1 Final certificate

contain many latent defects. Any amount agreed or notified, even if referred to as the final contract price, cannot be binding, but must be based on the assumption that work that appears to be compliant is in fact complaint, and adjusted following the end of the defects fixing period. The contract does not explain how that adjustment would be assessed, but if no agreement is reached, the contract administer should issue a revised notification of the final contract price promptly.

7.21   Clause 7.12 states: 'No later than 5 days after the end of the Defects Fixing Period or, where applicable, the period set out in clause 7.11, whichever is the later … the Architect/ Contract Administrator shall issue a final Payment Certificate (see Figure 7.1) which complies with the requirements of clause 7.5 representing the amount due for final payment' (cl. 7.12.1). If the contract administrator fails to issue the certificate, the contractor may issue a final payment notice (cl. 7.12.2). The certificate or payment notice should be calculated on the same basis as that for interim payments.

7.22   The client or contractor is required to pay the amount shown on the final payment certificate or notice. The 'payment due date' is the date of the certificate or notice, whichever is applicable (cl. 7.13) and, under clause 7.7, the final date for payment will be 14 days from the due date. The requirement to pay is, as with interim certificates, subject to the right to issue a pay less notice

Conclusiveness

7.23   Unlike in other contracts, it should be noted that no payment certificates are stated to be conclusive evidence that any matters or duties under the contract have been finally discharged, which means that even after the final payment certificate is issued, it is possible for either party to raise a claim for breach of contract. Having said that, it should be noted that if a matter is raised that could have been raised under the currency of the contract, and there is no good reason why the contractual mechanisms were not used to resolve it at that time, then it is unlikely that the claim would be successful.

8 Insurance

8.1   Death, injury of people and damage to property are all real possibilities in construction projects and so are risks which any contract needs to address. Almost all contracts, therefore, include provisions to deal with these.

8.2   Three interrelated concepts are usually used to cope with these particular risks: allocation of liability, indemnity and insurance. With respect to liability, if a party is made liable for a risk, it will normally bear the costs of the reasonably foreseeable consequences if that risk materialises. If a party agrees to indemnify another against a risk, it agrees to compensate that party for any losses it suffers should the risk occur. Finally, if a party is required to insure against a loss, the policy it takes out will cover it (and possibly the other party) for any losses.

8.3   The insurance clauses, including the numbering, are the same in the two versions of the RIBA Building Contract.

Liability

8.4   The RIBA Building Contracts allocate liability for insurable risks under clauses 6.1 and 6.2. (These are, of course, not the only risks that each party is liable for; risks are also allocated under, for example, cl. 9.3, which relate to delay.)

8.5   Clause 6.1 sets out the extent of the client's liability as follows:

In so far as the event has not been caused by the Contractor, or its employees or agents, in carrying out the Works, the Client's liability includes:

6.1.1 damage to existing structures and fixtures

6.1.2 damage to neighbouring property caused by the carrying out of the Works

8.6   The contractor's liability is set out in clause 6.2 as follows:

In so far as the event has not been caused by the Client, or its employees or agents, in carrying out the Works, the Contractor's liability includes

6.2.1 loss of or damage to the Works

6.2.2 loss of or damage to the property

6.2.3 loss of or damage to products and equipment

6.2.4 death of or bodily harm to any person working for the Contractor, arising in connection with the Works during the course of their employment

6.2.5 death of or bodily harm to a third party caused by the carrying out of the Works

6.2.6 any other loss or damage that is not the liability of the Client.

The contractor is therefore liable for the injury or death of any person engaged by it in relation to the works, and that of any third party that is caused by the works. It is also liable for loss or damage to the works, to 'the property' and to products or equipment, due to any cause.

8.7   Although this is generally clear, and similar to the allocation in other standard form contracts, it is worth noting some points. Under JCT contracts the employer is not made liable for the risks listed in clause 6.1.2 of the RIBA Building Contracts, i.e. for damage to neighbouring property not caused by the negligence of the contractor. A building owner may be liable to its neighbour under common law principles if work it carries out damages the neighbour's property, but this liability is not automatic, and the owner would not normally accept it under the building contract.

8.8   The contractor's liability under clause 6.2 is more widely defined than under the equivalent clauses in JCT contracts (e.g. IC16 cl. 6.1 or MW16 cl. 5.1). In JCT forms a contractor would only be liable for damage to property (other than the works) if it is negligent and if the damage arises out of the works; in clauses 6.2. these limitations do not apply. There is also some potential overlap between the allocation of liabilities (i.e. both parties being liable for the same losses). For example, the term 'the property' in clause 6.2.2 is not defined, and therefore could include the contractor's property, third-party property, the works, and the client's property, including the existing building (provided the damage is not caused by the client). However, under clause 6.1 the client accepts the risk of damage to existing structures and neighbouring property (if not caused by the contractor); what would therefore be the position if such damage occurs and neither party has caused it? It appears that both may be liable.

8.9   There is also potential overlap between clauses 6.2.3 and 6.1.1 (i.e. products and equipment that form part of existing fixtures) and between clauses 6.2.3 and 6.1.2 (i.e. products and equipment that constitute neighbouring property). It is probably intended that the 'products and equipment' are those belonging to the contractor and being used as part of the works, but the clause does not make this clear. In all these cases it is likely that, as the clause is clearly intended to distribute liability, a court would take a common sense approach and construe it in a way that would remove the overlaps. However, to avoid any arguments it might be sensible for the parties to introduce a clarification before entering into the contract.

Indemnity

8.10   It should be noted that there is no requirement for the contractor to indemnify the client against claims. So, if the client is sued by a party for losses caused by a matter that is the contractor's liability, it may have to settle the claim and then pursue the contractor for compensation. This is in contrast to the typical provisions in JCT contracts (e.g. IC16 cl. 6.1 and 6.2 and MW16 cl. 5.1 and 5.2), which require indemnification of the client against claims for injury to or death of persons, or damage to neighbouring property that has been caused by the contractor's negligence.

Insurance

8.11   The purpose of insurance is to ensure that those covered are compensated should the covered risks materialise. In the RIBA Building Contracts, as with most building contracts, the insurance provisions are linked to specific liabilities, and ideally the insurance policies should reflect precisely those liabilities. Even if the insurance is inadequate or non-existent, this would not affect or reduce a party's liability under the contract, or under the law, but in practice it would frequently not have adequate funds to compensate the other party, or affected third parties, for the losses. Any mismatch or lack of clarity in insurance coverage may therefore give rise to arguments, at a time when delays and complications will only exacerbate an already difficult situation.

8.12   Insurance is a complex and specialist subject area. The guidance given here is a brief explanation only, covering some of the key concepts. Which policies will be needed will depend on the particular circumstances of the project; clients are recommended to take specialist advice, and to include any specific requirements in the tender documents.

8.13   The contracts state that 'each Party is responsible for arranging insurance that is stated to be its responsibility in item J of the Contract Details for the amounts stated and for keeping it in place until Practical Completion' (cl. 6.3). It is important to note that the parties are *not* required to arrange insurance to cover their liabilities under clauses 6.1 and 6.2. Unlike other forms, there is no link between the liability clauses and the insurance clauses.

8.14   It is therefore very important that the parties give careful consideration as to whether the proposed insurance arrangements will cover them for the above risks, and set out full information on what insurance is required. The insurance clauses do not set out any details, but helpful advice is given in the guidance notes included in the form. Item J of the Contract Details requires the 'type', and 'amount' of insurance to be inserted, and indicates that insurance for damage to the works and to existing structures should be all risks and in joint names. It is suggested that, in addition, full information about the type of losses to be covered, exclusions, subrogation, etc. should also be given.

8.15   The contractor's liability for injury and death of employees (cl. 6.2.4) is met by its employers' liability insurance. This insurance is compulsory under the Employers' Liability (Compulsory Insurance) Act 1969. The legal minimum level of cover for most firms is £5 million, but many insurers will provide a £10 million policy as standard. Item J of the contract particulars indicates that the contractor is responsible for taking out this insurance, but an amount of cover must be inserted otherwise it is unclear what the extent of the obligation would be.

8.16   As noted in the contract guidance notes, the contractor's liability in respect of third parties (death or personal injury and loss or damage to property including consequential loss, covered under cl. 6.2.5) is met by its public liability policy. Insurers advocate insuring for a minimum of £2 million for any one occurrence, and insurance companies typically offer £5 million as the standard level of insurance. As above, item J of the contract particulars indicates that the contractor is responsible for this insurance, but an amount of cover must be inserted.

8.17   A contractor's public liability policy will not usually cover it for damage to neighbouring property, unless this was caused by its own negligence. Under the RIBA Building Contracts, this type of risk is accepted by the client, (cl. 6.1.2). The client will need to check if any existing building insurance it has will cover this risk; if it does not, the client needs to arrange for it to be extended, or take out a new policy. This insurance is usually expensive, and subject to a great many exclusions. The policy needs to be effective at the start of the site operations, when demolition, excavation, etc. are carried out. If the client would prefer the contractor to take out this insurance, the details will need to be given in the tender documents, and the clauses regarding liability and insurance amended accordingly, and an appropriate entry made in item J. (Under SBC16 this type of insurance is an optional provision, taken out by the contractor, and it is not included at all in MW16.)

8.18   With respect to the works, if the project is a new build, the contractor may be required take out a joint-names policy to cover any damage to the project (cl. 6.2.1, 6.2.3 and item J). With respect to existing buildings, the contractor is normally required to insure the works, and the client the existing structure (cl. 6.1.1, cl. 6.2.1 and cl. 6.2.3). This is similar to the approach in MW16 clause 5.4B. As noted above, full details would need to be inserted in item J, or the parties might find themselves liable for the losses under 6.1 and 6.2, but without adequate insurance cover. The arrangements should be discussed with both parties' insurance companies to ensure there are no gaps or overlap. As noted above, if the existing structure is damaged due to a fire caused by the contractor's negligence, is this to be part of the contractor's liability under clause 6.2.2 and, if so, is it to be covered by its public liability insurance (unlikely), or is it intended that the client's existing property insurance will cover such losses?

8.19   One issue that should be considered is that of subrogation, i.e. the right of an insurer to pursue a claim against a third party that caused a loss, in order to recover an amount paid to the insured for that loss. Any joint-names policy should make clear that, under the policy, the insurer does not have a right of subrogation to recover any of the monies from either of the named parties. In addition, the client should consider whether the policies should also include a waiver of any rights of subrogation against any subcontractors or required specialists (as they would in JCT contracts).

8.20   It is vitally important that all insurance matters are sorted out before the project starts on site. If there is a gap or an overlap (i.e. both parties insure against the same event or loss) this may cause serious difficulties. The last thing that is needed if a disaster such as a fire should occur is that the contractor or client is unable to honour its liabilities due to lack of funds, and the insurance companies become locked in a dispute and refuse to pay for the essential remedial work. The party responsible must provide evidence that adequate insurance has been taken out no later than 10 days before the start date (and any time after on request); if it is not provided the other party may take it out and the contract price is adjusted accordingly (cl. 6.4).

Professional indemnity insurance

8.21   Under optional clause 15, the contractor is required to ensure that there is adequate professional indemnity insurance for its design responsibilities (cl. 15.5). A professional indemnity policy insures a firm providing services against losses it suffers due to claims against it for negligence. This type of insurance protects the firm, but consequentially reduces the risk for the client. Should the building suffer defects due to negligent design, and the designer in question has no funds to cover the losses, there would be little purpose in bringing a claim against the designer. However, if an insurance policy is in place, the insurance company will compensate the client. Architects are required by their registration body to have professional indemnity insurance, but it cannot be assumed that a contractor will carry this.

9 Termination

9.1   Given the complexity and unpredictability of construction operations, it would be unlikely that a project could proceed to completion without breaches of the contractual terms by one or other of the parties. This is recognised by most construction contracts, which usually include provisions to deal with foreseeable situations. These provisions avoid arguments developing or the need to bring legal proceedings as the parties have agreed in advance machinery for dealing with the breach.

9.2   A clear example of this is the provisions for liquidated damages – the contractor is technically in breach if the project is not completed by the contractual date, but all the consequences and procedures for dealing with this are set out in the contract itself.

9.3   However, some breaches may have such significant consequences that the other party may prefer not to continue with the contract, and for these more serious breaches the contract will contain provisions for terminating the employment of the contractor.

9.4   In any contract, if unforeseen events mean that it becomes impossible for the contractual obligations to be fulfilled, the contract is sometimes said to be 'frustrated' and it may be set aside. In addition, where the behaviour of one party makes it difficult or impossible for the other to carry out its contractual obligations, the injured party might allege prevention of performance and sue either for damages or a *quantum meruit.* This could occur in construction where, for example, the client refuses to allow the contractor access to part of the site. Where it is impossible to expect further performance from a party, the injured party may claim that the contract has been repudiated. Repudiation occurs when one party makes it clear that it no longer intends to be bound by the provisions of the contract. This intention might be expressly stated, or it might be implied by the party's behaviour. In addition to these common law rights, construction contracts normally include express provisions allowing the parties to terminate the contractor's employment should serious situations occur, including but not necessarily limited to those that would amount to frustration or repudiation.

9.5   The idea of terminating the contractor's employment is something that most parties would seek to avoid if at all possible. Dealing with the consequences of termination, and the prospect of having to engage another contractor to complete a half-finished building, are difficult and stressful, and generally parties are better off trying to resolve their differences. However, sometimes the situation becomes so difficult that no other option is feasible. In such circumstances, the RIBA Building Contracts, like all other standard contracts, set out reasons and procedures for termination.

Termination by the client

9.6   The clauses relating to termination are identical in both versions of the contract. The client may terminate the contractor's employment under clause 12.2, with the reasons listed as giving grounds for termination by the client covered in clause 12.1. This states that the client may terminate the contractor's employment if the contractor:

- abandons the works;
- fails to proceed regularly and diligently;
- fails to comply with instructions;
- is in 'material breach' of the contract.

Abandoning the works

9.7   This would require that the contractor has made no appearance at the site for a significant period of time and has failed to satisfactorily explain why. An absence of a day or two would probably not be significant, but if there is no response to enquiries from the contract administrator, it would be sensible for the contract administrator to issue a notice of termination straight away.

Failing to proceed regularly and diligently

9.8   The phrase 'regularly and diligently' appears in many contracts, and has been the subject of much litigation. The client already has a remedy for slow progress and late completion, in the form of liquidated damages, so a generally poor performance is not normally considered sufficient to justify termination. The phrase means more than simply falling behind any submitted programme, even to such an extent that it is quite clear the project will finish considerably behind time. However, something less than a complete cessation of work on site would be sufficient grounds.

9.9   In the case of *London Borough of Hounslow* v *Twickenham Garden Developments* (1970), for example, the contract administrator's notice was strongly attacked by the defendants. In a more recent case, however, the contract administrator was found negligent because it failed to issue a notice. In *West Faulkner Associates* v *London Borough of Newham* (1992) the court stated:

> Taken together the obligation on the contractor is essentially to proceed continuously, industriously and efficiently with appropriate physical resources so as to progress the works steadily towards completion substantially in accordance with the contract requirements as to time, sequence and quality of work.

9.10   However, although failure to comply with a master programme would not by itself be a breach, it may be some evidence of failure to proceed regularly and diligently. This is where careful records will help to establish a case, and the regular updates to programmes and the records of discussions at progress meetings may be invaluable.

9.11   In any event, the right to terminate for failure to progress should not be used lightly. In particular, a client that realises the liquidated damages might not compensate it sufficiently should be advised that termination cannot be simply thought of as a convenient alternative. Generally, the failure has to be of reasonable significance, and here the programme would be of considerable help, particularly if it shows the resources to be deployed; if the resources actually deployed are substantially less than planned over a period of weeks, then a default could normally be established.

London Borough of Hounslow v *Twickenham Garden Developments* (1970) 7 BLR 81

The London Borough of Hounslow entered into a contract with Twickenham Garden Developments to carry out sub-structure works at Heston and Isleworth in Middlesex. The contract was on JCT63. Work on the contract stopped for approximately eight months due to a strike. After work resumed, the architects issued a notice of default stating that the contractor had failed to proceed regularly and diligently and that, unless there was an appreciable improvement, the contract would be determined. The employer then proceeded to determine the contractor's employment. The contractor disputed the validity of the notices and the determination, and refused to stop work and leave the site. The Borough applied to the court for an injunction to remove the contractor. The judge emphasised that an injunction was a serious remedy and that before he could grant one there had to be clear and indisputable evidence of the merits of the Borough's case. The evidence put before him, which showed a significant drop in the amounts of monthly certificates and numbers of workers on site, failed to provide this.

West Faulkner Associates v *London Borough of Newham* (1992) 61 BLR 81

West Faulkner Associates were architects engaged by the Borough for the refurbishment of a housing estate consisting of several blocks of flats. The residents of the estate were evacuated from their flats in stages to make way for the contractor, Moss, which, it had been agreed, would carry out the work according to a programme of phased possession and completion, with each block taking nine weeks. Moss fell behind the programme almost immediately. However, Moss had a large workforce on the site and continually promised to revise its programme and working methods to address the problems of lateness, poor quality work and unsafe working practices that were drawn to its attention on numerous occasions by the architect. In reality, Moss remained completely disorganised, and there was no apparent improvement. The architects took the advice of quantity surveyors that the grounds of failing to proceed regularly and diligently would be difficult to prove, and decided not to issue a notice. As a consequence, the Borough was unable to issue a notice of determination, had to negotiate a settlement with the contractor and dismissed the architect, which then brought a claim for its fees.

The judge decided that the architect was in breach of contract in failing to give proper consideration to the use of the determination provisions. In his judgment, he stated that 'regularly and diligently' should be construed together and in essence they mean simply that the contractors must go about their work in such a way as to achieve their contractual obligations. 'This requires them to plan their work, to lead and manage their workforce, to provide sufficient and proper materials and to employ competent tradesmen, so that the Works are carried out to an acceptable standard and that all time, sequence and other provisions are fulfilled' (Judge Newey at page 139).

Failing to comply with instructions

9.12   In JCT contracts, failure by the contractor to comply with just one instruction may be enough to terminate, but in the RIBA Building Contracts, as the plural is used, it appears something more is needed (cl. 12.1.3). This may well depend on what is involved. For example, if an instruction relates to health and safety, structural stability, security, compliance with legislation, etc., then failure to comply with repeated instructions relating to the same matter may be enough. Where it is more detailed or cosmetic, failure to comply with numerous instructions regarding a wide variety of matters may be needed.

Material breach of contract

9.13   The concept of a 'material breach' (cl. 12.1.4) giving a right to terminate is something that will be unfamiliar to those who normally use JCT contracts. Potentially, it is far wider in coverage than the more specific list of reasons to terminate given in, for example, MW16 (where the defaults include suspension without reasonable cause, failure to proceed regularly and diligently, and failure to comply with the CDM Regulations). However, it is frequently cited as a cause in other contracts, particularly bespoke ones. There are no absolute rules as to what types of breach would be considered 'material'; clearly they would need to be more than simply a trivial or a technical breach, but something less than a repudiatory breach.

9.14   So how serious and significant would a breach need to be to be considered material? Generally, it will depend on the nature of the project, the impact on the client and the circumstances surrounding the breach. The case of *SABIC UK Petrochemicals Ltd* v *Punj Lloyd Ltd* gives some guidance.

> *SABIC UK Petrochemicals Ltd* v *Punj Lloyd Ltd* [2013] EWHC 2916 (TCC), [2013] EWHC 3202 (TCC)
>
> Punj Lloyd was the Indian parent company of an insolvent contractor that had undertaken the construction of a low-density polyethylene plant on the old ICI site at Wilton for SABIC, the employer. SABIC terminated the contract for poor performance and commenced litigation. The court held SABIC's termination justified, although it did not amount to a repudiatory breach. It considered that there were aspects of the contractor's conduct that amounted to deliberate decisions not to comply with all of its contractual obligations, one of which was instructing its subcontractors to demobilise. However, the court's view was that, as at all material times the contractor stated its intention to bring the project to completion, that could not necessarily be equated with a renunciation of its side of the bargain. In that case, the contractor's conduct came close to being repudiatory but 'didn't cross the line'.

9.15   It is suggested that any of the following might constitute a material breach:

- refusal to comply with instructions;
- refusal to remove or correct defective work;
- failure to engage sufficient labour, removal of essential plant from the site;
- failure to comply with statutory obligations, particularly those regarding health and safety;
- any criminal act or evidence of corruption.

Procedure for terminating

9.16   The procedure for termination under clauses 12.1 and 12.2 follows a two-stage process. First, the contract administrator must issue the contractor with a notice of intention to terminate, referring to clause 12.1, and stating the reason for the termination. This is an essential first step, and any attempt to terminate the contractor's employment without it would be a breach of contract. Although not essential, it may be sensible for the contract administrator to set out exactly what would be needed for the default to be rectified.

9.17   All notices relating to termination are to be sent by recorded delivery as set out in clause 11.9, and are effective from the date of delivery. In addition, as this is a serious step that could ultimately bring the project to an end, it would be advisable for the contract administrator to discuss it with the client beforehand.

9.18   At the second stage, the contract administrator may issue a notice of termination (cl. 12.2). Before the notice can be issued, the contractor must have failed to remedy the default within 14 days of receiving the initial notice of intention to terminate. It would, of course, be essential that the contract administrator discusses this with the client before taking this step, as essentially termination of the contract should be a matter for the client to decide.

Termination by the contractor

9.19   The contractor may terminate its employment under clause 12.4. The grounds for termination by the contractor are covered in clause 12.3, which states:

> If the Client is in Material Breach of the Contract, then the Contractor shall issue the Client with a 14-day notice of its intention to terminate, referring to this clause and stating the Material Breach.

Material breach of contract

9.20   As with termination by the client, a breach would need to be more than merely trivial, but not as significant as a full repudiatory breach. It is suggested that any of the following might constitute a material breach by the client:

- refusal to allow access to the site or parts of the site;
- failure to comply with statutory obligations, particularly those regarding health and safety;
- failure to supply information essential for completion of the works;
- employing others to carry out the works;
- seeking to terminate the contractor's employment on grounds not allowed under the contract.

9.21   A notable change since the last edition of the form is the removal of the ground 'failure to pay the contractor when payment is due'. In order for the contractor to terminate for failure to pay, the failure would have to be significant enough to amount to a material breach.

9.22   The procedure for termination follows a similar two-step process, i.e. first the contractor issues a 14-day notice then, if the situation is not resolved, it may terminate its employment by means of a further notice (cl. 12.4).

Termination by either party

9.23   Either party may terminate the contractor's employment due to insolvency, bankruptcy or frustration.

Insolvency or bankruptcy

9.24   Clause 12.5 states:

> If a Party is declared insolvent or bankrupt under any applicable law, the other Party may terminate the contractor's employment by issuing the insolvent Party with a notice of termination.

9.25   Unlike JCT contracts, no definition of insolvency is given. This may be sensible as the definition under the law may change over time, and this is a matter of statute rather than compliance with contractual provisions. However, as a guide the parties could consult the current definition in a JCT contract (e.g. the IC16, cl. 8.1) or other up-to-date text. If there is any doubt about the matter, the client should seek expert advice, as this is an issue that could have serious consequences.

9.26   It should be noted that, unlike in many of the older versions of JCT contracts, termination is not automatic upon insolvency or bankruptcy, and positive action is required by the other party to bring the contractor's employment to an end. However, in this case it is a single-step process, and only one notice is required to effect the termination. No time limits are given, which is sensible as it may be some time before the situation is clear. In some cases the parties may wish to continue to work together for some time in order to complete as much of the project as possible.

Frustration

9.27   Clause 12.6 states:

> Either Party may terminate the Contractor's employment if any event not caused by and not the responsibly of the Parties prevents the Works from being carried out for a continuous period of 60 days.

9.28   The clause is broadly worded, and it is suggested that 'any event not caused by (and not the responsibility of)' should be interpreted in a practical way. For example, the parties have assumed liability for various matters under clauses 6.1 and 6.2 (both contracts) that may not have been caused by them, such as loss of or damage to the works (cl. 6.2.1), and similarly the client has accepted various risks under clauses 9.1 to 9.3, such as force majeure and the actions of a utility company. It could be argued that having accepted liability the party is 'responsible' for the consequences. However, it is quite possible that these events might result in work ceasing for 60 days, and it is suggested that, provided a party did not actually cause the problem, it ought to be able to terminate the contractor's employment should this occur, even though under the contract it accepted liability for or the risk of this event.

9.29   No procedure is set out for terminating under this ground. For practical reasons some form of notice would obviously be needed, and it may be sensible to also issue a warning notice, following the procedure set out in clauses 12.1–12.4; however, there would be no obligation to do so.

Consequences of termination

9.30   The consequences of termination are set out in clauses 12.7–12.10. The clauses are primarily concerned with payment. It should be noted that in all cases the termination clauses refer to terminating the contractor's employment, not to terminating the contract therefore any clause referring to the consequences of termination or the liability of either party would still apply, unless the contract provides otherwise.

Payment

9.31   The contracts explain how the balance due is to be calculated in situations where the client has terminated the contractor's employment, and in situations where the contractor initiated the termination (cl. 12.7). They also establish when the balance is to become due (cl. 12.8 and 12.9).

9.32   If the client terminates the contractor's employment, the client is entitled to the costs for completing the work with another contractor, and any other reasonable costs consequent upon the termination (cl. 12.7.1). The balance is not due until all the work has been completed (cl. 12.8).

9.33   If the contractor initiated the termination it is entitled to all costs and losses incurred. The balance is due within 14 days of the contractor submitting an application for payment, or of the contract administrator issuing a payment certificate, whichever occurs first (cl. 12.9).

9.34   It should be noted that any amount which may have become due by the time of the termination does not need to be paid in situations where the client has issued a pay less notice, or if the contractor has become insolvent before the final date for payment (this reflects the position as set out by the House of Lords in *Melville Dundas* v *George Wimpey*).

> *Melville Dundas Ltd* v *George Wimpey UK Ltd* [2007] 1 WLR 1136 (HL)
>
> On a contract let on WCD98, the contractor had gone into receivership, entitling the employer to determine the contractor's employment. The contractor had applied for an interim payment on 2 May 2003, the final date for payment was 16 May (14 days after application), and the determination was effective on 30 May. The contractor claimed the payment on the basis that no withholding notice had been issued. By a majority of three to two, the House of Lords decided that the employer was not obliged to make any further payment. It was accepted that, under WCD98, interim payments were not contractually payable after determination and the House of Lords held that this was not inconsistent with the payment provisions of the HGCRA 1996. Although the Act requires that the contractor should be entitled to payment in the absence of a notice, this did not mean that that entitlement had to be maintained after the contractor had become insolvent, i.e. it was not inconsistent to construe that the effect of the determination was that the payment was no longer due. The Act was concerned with the balance of interests between payer and payee, and to construe it otherwise would give a benefit to the contractor's creditors against the interests of the employer, something which the Act did not intend.

Access to the site and security

9.35   If the contractor's employment under the contract is terminated for any reason, clause 12.10 states that the contractor shall:

- lose its right to access the site;

- remove all its materials and equipment within a reasonable time;

- no longer be responsible for the security of the site.

9.36   This is a sensible clause, which makes the position clear to the parties should termination occur. The client should note that it will immediately become responsible for the site, and will probably need to take steps to ensure that it is secure and complies with any health and safety or other statutory requirements. The client will also need to allow the contractor access to remove its materials and equipment; the contractor should give reasonable notice of when it intends to do this so that arrangements can be made.

10 Dispute handling and resolution

10.1 At the start of a project the idea that the parties may fall into dispute may seem only a remote possibility and therefore not worth considering in detail. However, the reality is that once a dispute has arisen it will not be possible to sensibly agree a way forward to its resolution. It is therefore very important that care is taken at the outset to select dispute resolution options that are appropriate and with which both parties are happy.

10.2 Ideally, if a dispute breaks out it should first be handled informally, through negotiation, correspondence or, perhaps, at a specially called meeting. The advance warning notices required under clause 3.2, and any related meetings, will also help to avert or resolve many matters. Embarking on more formal dispute resolution methods is not something to be undertaken lightly; they will all involve the parties in additional costs and time (although to varying degrees) and may be very stressful. In addition, with the exception of mediation, they are likely to result in a poor relationship between the parties for the remainder of the project.

10.3 However, in some cases the need for a more formal method of dispute resolution will be unavoidable. For such situations, the RIBA Building Contracts offer several options: mediation, adjudication, arbitration and litigation. The parties are required to decide on the methods to be used before entering the contract, by means of entries in item N of the Contract Details.

10.4 In DBC, mediation, adjudication and arbitration can be selected, or none, or any combination. Litigation will apply by default as the final method of dispute resolution if arbitration is not selected. In CBC the adjudication option is preselected (see paras 2.24 and 2.25), but otherwise the same applies. The parties are invited to name the arbitrator, adjudicator or mediator of their choice, but if none is named, the guidance note to item N states that 'selection can be made by the Royal Institute of British Architects'.

10.5 A brief outline of each method is set out below, in order to highlight the key differences between the procedures. More information can be found in the many texts on these topics, including the *RIBA Good Practice Guides* on mediation, arbitration and adjudication.

Mediation

10.6 Mediation is a voluntary process whereby the parties are helped by a professional mediator to resolve the dispute in a way that they are both comfortable with. The mediator has no authority to impose a solution, and it may be that the dispute is not resolved. However, as it is less adversarial than other methods, even if a solution is not reached the mediation may have helped to pave the way for further discussions. It is also less expensive than other methods, particularly if the mediation is limited to one day. The parties should bear in mind that they will nevertheless need to pay their own expenses, the mediator's fee and the cost of the venue.

10.7   If this option is selected, the mediation would normally happen before any other steps are taken. It is not stated to be a 'condition precedent', so if it is selected, there would be nothing to prevent either party initiating adjudication, arbitration or court proceedings at the same time. However, if the parties are genuinely interested in resolving the dispute they are unlikely to take such steps while the mediation is in progress.

10.8   The RIBA offers a mediation scheme, whereby the mediator will be appointed to work with the parties, and is normally paid on a daily rate.[1] For a small project, the mediation may last for a day or even less, but for large, complex disputes it could take several days. There are no set rules for mediation, and each mediator will have various techniques that they may use, for example meeting the parties separately, taking issues and suggestions back and forward between the two, and/or bringing them together for a chaired discussion. All proceedings are confidential and on a 'without prejudice' basis, so that offers made cannot be raised in future dispute proceedings, and nothing disclosed to the mediator will be disclosed to the other party unless a party permits it.

10.9   At the end of the mediation, if an agreement is reached, the parties will sign a binding agreement, which is enforceable in the same way that any contract between the parties would be. If the parties are still in deadlock they can ask the mediator to propose a solution, but of course they do not have to accept it.

Adjudication

10.10   Adjudication is a statutory right of any party to a construction contract that falls under the definition in the Housing Grants Act (see para. 1.14). It is a procedure by which the dispute may be referred by either party to an adjudicator, who must reach a decision within 28 days. In construction industry terms, this is a relatively short period, although it can be extended by 14 days by the referring party, and further by agreement. In the case of projects with a residential occupier (which would be the case with most projects under DBC, see Appendix 1: para A1.7), the Housing Grants Act does not apply by default, therefore adjudication would need to be selected in item N of the Contract Details if it is to be used.

10.11   Adjudication has advantages and disadvantages, the key advantage being that the matter is sorted out quickly, which limits the amount of time and resources that the parties can spend on it. The disadvantage is that the whole process can seem extremely rushed, particularly from the client's perspective. Usually, the dispute is initiated by the contractor, who may have spent a considerable period in advance preparing for the process, but the client may be required to respond very quickly to what can be a lengthy claim. The decision may seem rather 'rough justice' as there is little time for the responding party to develop and put forward careful arguments, or to commission technical reports and collect other relevant evidence. On the plus side, the decision, although it must be complied with, can be challenged in the sense that the dispute can later be raised in arbitration or in court.

---

[1]   Details can be downloaded at www.architecture.com/-/media/gathercontent/dispute-resolution/additional-documents/fixedfeemediationguidancepdf.pdf

The adjudication process

10.12   Neither version of the RIBA Building Contract sets out a detailed procedure for adjudication. Instead, the contracts state that the rules will be as stated in item N of the Contract Details (cl. 13.6, CBC; cl. 13.2.5, DBC). In the case of CBC, this refers to secondary legislation, the Scheme for Construction Contracts[2] (the Scheme, see para. A1.6), whereas DBC refers to the RIBA Adjudication Scheme for Consumer Contracts (the RIBA Scheme). The most significant difference between these is that the latter is a shorter procedure, and the adjudicator's fees are capped.

10.13   Under the statutory Scheme, the party wishing to refer a dispute to adjudication must first give notice to the other party identifying briefly the dispute or difference, giving details of where and when it has arisen and setting out the nature of the redress sought. If no adjudicator is named, the parties may either agree an adjudicator or either party may apply to any nominating body specified in the contract or, if none is specified, to any nominating body. CBC states that 'selection can be made by the RIBA' but as this is arguably not a binding agreement to apply only to this body, it might be sensible for the parties to confirm the agreed nominating body is clearly under item N. Once appointed, the adjudicator will normally then send terms of appointment to the parties. In the case of the RIBA Scheme, the party wishing to have the dispute resolved by adjudication must apply to the RIBA.

10.14   Under the statutory Scheme, the referring party must refer the dispute to the selected adjudicator within 7 days of the date of the notice (Scheme para. 7(1)). The referral will normally include particulars of the dispute, and must include a copy of, or relevant extracts from, the contract, and any material the party wishes the adjudicator to consider (para. 7(2)). A copy of the referral must be sent to the other party, and the adjudicator must inform all parties of the date it was received (para. 7(3)). The adjudicator will then set out the procedure to be followed. A preliminary meeting may be held to discuss this, otherwise the adjudicator will send the procedure and timetable to both parties. The party that did not initiate the adjudication (the responding party) will be required to respond by a stipulated deadline. The adjudicator may hold a short hearing and/or may visit the site. Occasionally, it may be possible to carry out the whole process by correspondence (often termed 'documents only'). The adjudicator must reach a decision within 28 days of the referral, unless extended by agreement.

10.15   The RIBA Scheme does not set out any particular procedure. It is left entirely to the discretion of the adjudicator to direct the parties as to whether and when referral documents and a response are to be submitted, but in most cases the adjudicator will ask for information from each side and may, as above, decide to hold a meeting. In this case the decision must be reached within 21 days of the adjudicator being appointed.

10.16   The adjudicator is required to act impartially (Scheme para. 12(a); RIBA Scheme para. 16). The statutory Scheme states that the adjudicator is not liable for anything done or omitted when acting properly as an adjudicator (para. 26).

10.17   Under the statutory Scheme the parties must meet their own costs of the adjudication, unless they have agreed that the adjudicator shall have the power to award costs, which

---

2   See www.blplaw.com/media/pdfs/Statutory_Instrument_1998_No._649_2.pdf

they may only do after the dispute has arisen. The adjudicator, however, is entitled to charge fees and expenses (subject to any agreement to the contrary) and may apportion those fees between the parties. The parties are jointly and severally liable to the adjudicator for any sum that remains outstanding following the adjudicator's determination. This means that in the event of default by one party, the other party becomes liable to the adjudicator for the outstanding amount. In the case of the RIBA Scheme the parties must meet their own costs. The adjudicator's fees are limited to £150 per hour (exclusive of VAT) up to a maximum of fifteen hours, and the adjudicator may apportion those fees between the parties. The parties are not jointly and severally liable for the fees, instead the adjudicator may take court proceedings against any party that does not pay its apportioned amount.

Challenging an adjudicator's decision

10.18   The Scheme states that the adjudicator's decision will be final and binding on the parties 'until the dispute is finally determined by legal proceedings, by arbitration (if the contract provides for arbitration or the parties otherwise agree to arbitration) or by agreement between the parties' (para. 23(2)). The effect of this is that if either party is dissatisfied with the decision, it may raise the dispute again in arbitration or litigation as indicated in the Contract Details, or it may negotiate a fresh agreement with the other party. In all cases, however, the parties remain bound by the decision and must comply with it until the final outcome is determined.

10.19   The RIBA Scheme states that 'The customer and the contractor must follow the adjudicator's decision as part of their obligations under the building contract, unless and until either party obtains a court judgment about the dispute which is different from the decision of the adjudicator' (para. 22). There is no reference to final determination by arbitration, possibly because the RIBA Scheme was originally developed to cater for the JCT Building Contracts for a Home/Owner Occupier, which do not include arbitration as a dispute resolution option. Users of a RIBA Domestic Building Contract that intend to select arbitration as the final forum may wish to include an appropriate amendment to paragraph 22 of the RIBA Scheme under item P.

10.20   If either party refuses to comply with the decision, the other may seek to enforce it through the courts. Generally, actions regarding adjudicators' decisions have been dealt with promptly by the courts and the recalcitrant party has been required to comply. Paragraph 22A of the Scheme allows the adjudicator to correct clerical or typographical errors in the decision, within 5 days of it being issued, either on the adjudicator's own initiative or because the parties have requested it, but this would not extend to reconsidering the substance of the dispute.

Arbitration

10.21   Arbitration is essentially a private alternative to court proceedings. If arbitration is not selected, the default process for the ultimate resolution of disputes will be the courts (cl. 13.9, CBC; cl. 13.5, DBC). If it is selected, either party has the right to require that any dispute is taken to arbitration (cl. 13.8, CBC; cl. 13.4, DBC). If one of the parties nevertheless initiates court proceedings, the other can apply for a stay, and the courts would normally freeze the proceedings to allow the arbitration to take place.

10.22 The process is supported by statute (the Arbitration Act 1996), and a court would enforce the decision ('award') of an arbitrator in the same way that it would enforce its own judgments. There are very limited rights to challenge the enforcement of an award, the grounds of which are confined to matters such as lack of jurisdiction of the arbitrator or bias. In some circumstances a party may appeal the award on a point of law, but even this right can be excluded if the parties agree.

10.23 Arbitration is a private process, which is often an attractive feature to clients and their consultants. It is also very flexible, as the parties can agree the timetable and venue; failing agreement, the arbitrator has authority to direct these matters. The form of proceedings can vary hugely, but they are often relatively long and formal, and consequently expensive in comparison with mediation and adjudication. Various sets of rules exist that the parties can adopt, such as the Construction Industry Model Arbitration Rules (CIMAR).[3] These include rules for a 'documents only' arbitration (i.e. with no hearing, see Rule 8), for a short hearing (Rule 7) and for a full procedure (Rule 9).

10.24 Under the Arbitration Act the arbitrator has the power to award costs, unless the parties agree otherwise. Where the arbitrator has the power to award costs, this will normally be done on a judicial basis, i.e. the loser will pay the winner's costs (CIMAR Rule 13.1). The arbitrator will be entitled to charge fees and expenses and will apportion those fees between the parties on the same basis. The parties are jointly and severally liable to the arbitrator for fees and expenses incurred.

10.25 As the costs in an arbitration are often significant (sometimes amounting to more than the actual amount claimed), the issue of who pays them is of considerable concern to the parties. In an effort to reduce their liability for costs, the party that is being claimed against may make an offer to settle. If this is done correctly, and the other party refuses the offer but ultimately is awarded less than the sum offered, it will not be able to recover any of its costs from the date the offer was refused.

---

[3] These can be downloaded at www.jctltd.co.uk/docs/JCT_CIMAR_2016.pdf

Appendix 1

Key legislation that impacts on construction contracts

Sale of Goods Act 1979

Applies to: contracts for the sale of goods

A1.1   This implies terms into contracts for the sale of goods regarding title (section 12), correspondence with description (section 13), quality and fitness for purpose (section 14) and sale by sample (section 15). For the purposes of the Act, goods are of satisfactory quality if they meet the standard that a reasonable person would regard as satisfactory, taking account of any description of the goods, the price (if relevant) and all the other relevant circumstances. Section 14 implies a term that where the seller sells goods in the course of business and the buyer, expressly or by implication, makes known to the seller any particular purpose for which the goods are being bought, there is an implied condition that the goods supplied under the contract will be reasonably fit for that purpose.

Supply of Goods and Services Act 1982

A1.2   This covers contracts for work and materials, contracts for the hire of goods and contracts for services. Most construction contracts come under the category of 'work and materials' and the Act implies terms into these equivalent to sections 12 to 15 listed in paragraph A1.1 with respect to any goods in which the property has been transferred under the contract. So, as above, any materials supplied should be reasonably fit for their intended purpose, provided always that the buyer is relying on the supplier's skill and judgement. (If the buyer specifies a particular material then this would be sufficient to show that it was not relying on the seller.) For services, the Act implies terms regarding care and skill, time of performance and consideration. For example, section 14 implies a term that where the supplier is acting in the course of business, the supplier will carry out the services within a reasonable time, provided of course the parties have not themselves agreed terms regarding time.

Contracts (Rights of Third Parties) Act 1999

A1.3   This Act entitles third parties to enforce a right under a contract, where the term in question was intended to provide a benefit to that third party. The third party could be specifically named or it could be an identified class of people. For example, if a contract to execute work on a property named a prospective future tenant as a beneficiary of that contract, the tenant may be able to bring a claim if the work is done incorrectly. The Act, however, allows for parties to agree that their contract will not be subject to its provisions.

Defective Premises Act 1972

A1.4   This applies where work is carried out in connection with a dwelling, including design work. It states that 'A person taking on work for or in connection with the provision of a dwelling … owes a duty … to see that the work which he takes on is done in a workmanlike or, as the case may be, professional manner, with proper materials and so that as regards that work the dwelling will be fit for habitation when completed' (section 1(1)). This appears to be a strict liability, and is owed to anyone acquiring an interest in the dwelling.

Housing Grants, Construction and Regeneration Act 1996

Applies to: construction contracts, except where one of the parties is a 'residential occupier'

A1.5   The Housing Grants Act 1996 applies to most construction contracts (the term 'construction contract' is given a wide definition, and includes contracts for services only or for building work, and applies to a large range of types of work, including demolition works, services installations and repair work, as well as new buildings, extensions and alterations). The Act requires that construction contracts include specific terms relating to adjudication and payment:

- the right to stage payments;

- the right to notice of the amount to be paid;

- the right to suspend work for non-payment;

- the right to take any dispute arising out of the contract to adjudication.

A1.6   If the parties fail to include these provisions in their contract, the Act will imply terms to provide these rights (section 114) by means of the Scheme for Construction Contracts (England and Wales) Regulations 1998.

A1.7   However, there is an important exception; it does not apply to projects where one of the parties is a 'residential occupier'. The residential occupier exception applies to projects for which the primary purpose is beneficial use by the client as a residence (section 106). This would include buildings that the client is occupying or intending to occupy as its main residence, and might also include a second home if the client is the main user and there is no intention to use it as a holiday let (*Westfields Construction Ltd* v *Lewis*). However, work to buildings in the grounds of a residence that will not be lived in by the client, or work to divide a property into flats where only one flat will be retained by the client, will not fall under the exception (*Samuel Thomas Construction* v *Anon*). Similarly, work on other residential properties, for example for landlords, local authorities or housing associations, will usually be covered by the Act.

Westfields Construction Ltd v *Lewis* [2013] BLR 233 (TCC)

This case related to the enforcement of an adjudicator decision. A key question was whether 'residential occupation' should be assessed at a single point in time (e.g. at the time the contract is formed) or as an ongoing process. The judge was of the view that 'occupies' is an ongoing process and must carry with it some reflection of the future: it indicates that the employer occupies and will remain at (or intends to return to) the property. The employer had moved out of the property and had talked about letting it. The judge found on the evidence that the employer intended to let out the property, and the contract therefore did not fall within the 'residential occupier' exception.

Samuel Thomas Construction v *Anon* (unreported) 28 January 2000 TCC

This case also related to the enforcement of an adjudicator decision. The contract was not on a standard form, and the judge had to decide whether the section 106 'residential occupier' exclusion applied. If it did, the decision against the employer would be unenforceable. The contract concerned a number of buildings that were being refurbished, including a barn that the employer intended to occupy, and another barn and associated buildings that were being refurbished for sale. There was only one contract for the works. The judge upheld the adjudicator's view that where one dwelling was to be occupied and the other was not, the contract did not 'principally relate to operations on a dwelling which one of the parties ... intended to occupy' and therefore the exception did not apply.

Exemption clauses

A1.8   The scope for excluding liability for important matters is limited by two significant pieces of legislation, the Unfair Contract Terms Act 1977 and the Consumer Rights Act 2015.

Unfair Contract Terms Act 1977

Applies to: only to B2B (or C2C), not business to consumer

A1.9   This has the effect of rendering various exclusion clauses void, including any clauses excluding liability for death or personal injury resulting from negligence, any clauses attempting to exclude liability for Sale of Goods Act 1979 section 12 obligations (and the equivalent under the Supply of Goods and Services Act 1982), and any clauses attempting to exclude liability for Sale of Goods Act 1979 section 13, 14 or 15 obligations (and the equivalent under the Supply of Goods and Services Act) where they are operating against any person dealing as consumer. It also renders certain other exclusion clauses void in so far as they fail to satisfy a test of reasonableness; for example, liability for negligence other than liability for death or personal injury, and liability for breach of section 13, 14 or 15 obligations in contracts which do not involve a consumer.

Consumer Rights Act 2015

Applies to: only to B2C

A1.10   This Act, which came into force on 1 October 2015, consolidates much of the pre-existing legislation on consumer protection, and introduced some significant new provisions. It replaced the sections of the Unfair Contract Terms Act 1977 that relate to consumers, and repealed the Unfair Terms in Consumer Contracts Regulations 1999. It applies to contracts and notices between a 'trader' and a 'consumer'. A 'consumer' is defined as 'an individual acting for purposes that are wholly or mainly outside that individual's trade, business, craft or profession' (section 2(3)).

A1.11   The Act applies to a wider range of contracts than other legislation commonly encountered by construction professionals. For example, under the Housing Grants Act 1996 (as amended) only contracts with a residential occupier are excluded. It is therefore quite possible that the Consumer Rights Act will apply to a contract that is excluded from the Housing Grants Act 1996 (e.g. where an individual undertakes work to a domestic property that is not the individual's main residence).

A1.12   The Act states that any contract for services is to be treated as including a term that the trader must perform the service with reasonable care and skill (section 49(1)). In addition, if the contract does not provide for a price or timescale, it is taken to include a term that the services will be provided for a reasonable price (section 51), and within a reasonable timescale (section 52). Goods supplied under such a contract must also be of good quality.

A1.13   Part 2 sets out the law regarding unfair terms in relation to consumers. Section 62(1) states that an unfair term of a consumer contract is not binding on the consumer. The test for 'unfair terms' in the Act is the same as that in the 1977 Unfair Contract Terms Act: it provides that a 'term is unfair if, contrary to the requirement of good faith, it causes a significant imbalance in the parties' rights and obligations under the contract to the detriment of the consumer' (section 61(4)). An 'indicative and non-exhaustive list' of examples of what might be considered unfair terms is set out in Schedule 2 to the Act. This includes, for example, any term which has the object or effect of excluding or hindering the consumer's right to take legal action or exercise any other legal remedy, which would include an arbitration agreement.

A1.14   The most significant change relates to terms specifying the main subject matter of the contract or setting the price. These terms are *not* subject to the 'fairness' test provided that they are both transparent and prominent (section 64(1) and (2)). 'Transparent' is defined as being 'in plain and intelligible language and (in the case of a written term) is legible' (section 64(3)) and 'prominent' as 'brought to the consumer's attention in such a way that an average consumer [who is reasonably well-informed, observant and circumspect] would be aware of the term' (section 64(4) and (5)). The new Act no longer makes an exception for terms that have been 'individually negotiated' as had been the case in the Unfair Terms in Consumer Contracts Regulations 1999.

Other consumer legislation

Unfair Terms in Consumer Contracts Regulations 1999

A1.15   Repealed – entirely replaced by the Consumer Rights Act 2015 for any contract entered into after 1 October 2015. Essentially, these provided that terms that significantly affect the balance of rights between the parties, and which are not individually negotiated, may be deemed unfair and therefore void.

References

Publications

Birkby, G, Ponte, A and Alderson, F. *Good Practice Guide: Extensions of Time*, RIBA Publishing, London (2008).

Construction Industry Council. *Construction (Design and Management) Regulations 2015*. Risk Management Briefing. CIC, London (2015).

Coombes Davies, M. *Good Practice Guide: Adjudication*, RIBA Publishing, London (2011).

Coombes Davies, M. *Good Practice Guide: Arbitration*, RIBA Publishing, London (2011).

Eggleston, B. *Liquidated Damages and Extensions of Time: In Construction Contracts*, 3rd edition, Sweet & Maxwell, London (2009).

Finch, R. *NBS Guide to Tendering: For Construction Projects*. RIBA Publishing, London (2011).

Furst, S. and Ramsey, V. (eds.). *Keating on Construction Contracts*, 9th edition, Sweet & Maxwell,London (2012).

Grossman, A. *Good Practice Guide: Mediation*, RIBA Publishing, London (2009).

Health and Safety Executive. *Managing Health and Safety in Construction*, HSE Books (2015).

JCT. *Tendering Practice Note 2017*. Sweet & Maxwell, London (2017).

NBS National Construction Contracts and Law Report 2018 (RIBA Enterprises Ltd, 2018)

Ostime, N. *RIBA Job Book*, 9th edition, RIBA Publishing, London (2013).

Cases

BFI Group of Companies Ltd v *DCB Integration Systems Ltd* [1987] CILL 348	2.22
Cavendish Square Holdings v *El Makdessi* Supreme Court [2015] UKSA 67	2.22
City Inn Ltd v *Shepherd Construction Ltd* [2008] CILL 2537 Outer House Court of Session	4.23
Dhamija v *Sunningdale Joineries Ltd, Lewandowski Willcox Ltd, McBains Cooper Consulting Ltd* [2010] EWHC 2396 (TCC)	6.27
Domsalla v *Dyason* [2007] BLR 348	1.23
Gloucestershire County Council v *Richardson* [1969] 1 AC 480 HL	3.30
Goldsworthy and others v *Harrison and another* [2016] EWHC 1589 (TCC)	2.35
H W Nevill (Sunblest) Ltd v *William Press & Son Ltd* (1981) 20 BLR 78	4.3, 7.3
Hamid v *Francis Bradshaw Partnership* [2013] EWCA Civ 470	2.37
Henry Boot Construction (UK) Ltd v *Malmaison Hotel (Manchester) Ltd* (1999) 70 Con LR 32 (TCC)	4.39
Impresa Castelli SpA v *Cola Holdings Ltd* (2002) CLJ 45	4.3, 7.12
Jameson v *Simon* (1899) 1 F (Ct of Session) 1211	5.19
London Borough of Hounslow v *Twickenham Gardens Development* (1970) 78 BLR 89	4.3, 9.9
Lovell Projects v *Legg & Carver* [2003] BLR 452	2.25
McGlinn v *Waltham Contractors Ltd* [2007] 111 Con LR 1	5.18
Melville Dundas Ltd v *George Wimpey UK Ltd* [2007] 1 WLR 1136 (HL)	9.35
MT Højgaard A/S v *E.ON Climate & Renewables UK Robin Rigg East Ltd & Ors*	3.21
Mul v *Hutton Construction Limited* [2014] EWHC 1797 (TCC)	6.11
National Museums and Galleries on Merseyside (Trustees of) v *AEW Architects and Designers Ltd* [2013] EWHC 2403 (TCC)	3.15, 5.11
Parking Eye Limited v *Beavis, Supreme Court* [2015] UKSA 67	2.22

Legislation

Statutes

Statutory instruments

Further Reading

Chapter 1

Lupton, S and Stellakis, M. *Which Contract?: Choosing the Appropriate Building contract* (6th ed.), RIBA Publishing, London (to be published in 2019).

Chapter 2

Adriaanse, J. *Construction Contract Law: The Essentials* (4th ed.), Palgrave, London (2016).
CIC. *Building Information Modelling (BIM) Protocol* (2nd ed.), Construction Industry Council, London (2018).
JCT. *Tendering Practice Note 2017*, Joint Contracts Tribunal, London (2017).

Chapter 4

Burr, A. *Delay and Disruption in Construction Contracts* (5th Ed.), Informa Law from Routledge, London (2016).
Haidar, A. and Barnes, P. T. *Delay and Disruption Claims in Construction: A Practical Approach* (2nd ed.), ICE Publishing, London (2014).

Chapter 5

Bussey, P. *CDM 2015: A Practical Guide for Architects and Designers*, RIBA Publishing, London (2015).
Association for Project Safety, *Principal Designer's Handbook 2015: Guide to the CDM Regulations 2015*, RIBA Publishing, London (2016).

Chapter 6

Haidar, A. and Barnes, P. T. *Delay and Disruption Claims in Construction: A Practical Approach*, ICE Publishing, London (2014).
Whitfield, J. *RIBA Good Practice Guide: Assessing Loss and Expense*, RIBA Publishing, London (2013).

Chapter 7

Furst, S and Ramsey, V, *Keating on Construction Contracts* (10th ed.), Sweet & Maxwell, London (2016)

Chapter 8

Hogarth, R, Anderson, A. and Goldring, S. (eds) *Insurance Law for the Construction Industry*
    (2nd ed.), OUP Oxford, Oxford (2013).

Chapter 10

Coombes Davies, M. *Good Practice Guide: Adjudication*, RIBA Publishing, London (2011).
Coombes Davies, M. *Good Practice Guide: Arbitration*, RIBA Publishing, London (2011).
Grossman, A. *Good Practice Guide: Mediation*, RIBA Publishing, London (2009).
Kavanagh, B. *Avoiding and Resolving Disputes: A Short Guide for Architects*, RIBA Publishing,
    London (2017).

Clause Index *by paragraph number*

CBC cl.	DBC cl.	paragraph	CBC cl	DBC cl.	paragraph
5.8.3	5.8.3	5.30	7.15	7.15	6.47
5.8.4	5.8.4	5.30, 6.31	7.16	7.16	6.55
5.9	5.9	2.44	8.1	8.1	6.57
5.11	5.11	2.45, 5.31, 6.6, 6.12	8.2	8.2	6.59
5.12	5.12	5.31, 6.13	9.3	9.3	4.27, 4.30, 4.38, 8.4
5.13	5.13	5.31, 6.13	9.3.3	9.3.3	4.28
5.14	5.14	5.29	9.4	9.4	4.23
5.14.2	5.14.2	3.33	9.5	9.5	4.30
5.15	5.15	5.29	9.6	9.6	4.30, 4.35
6.1	6.1	8.4–8.9	9.7	9.7	6.6, 6.14–6.15, 6.30
6.1.1	6.1.1	8.5, 8.9, 8.18	9.9	9.9	6.15
6.1.2	6.1.2	8.5, 8.7, 8.9, 8.17	9.10	9.10	7.1
6.2	6.2	8.4, 8.6–8.8	9.10.1	9.10.1	7.4
6.2.1	6.2.1	8.6, 8.18, 9.28	9.11	9.11	7.5
6.2.2	6.2.2	8.6, 8.8, 8.18	9.11.1	9.11.1	7.5
6.2.3	6.2.3	8.6, 8.9, 8.18	9.11.2	9.11.2	7.7
6.2.4	6.2.4	8.6, 8.15	9.12	9.12	7.9, 7.11
6.2.5	6.2.5	8.6, 8.16	9.13.1	9.13.1	7.9
6.2.6	6.2.6	8.6	9.13.2	9.13.2	7.10
6.3	6.3	8.13	9.13.3	9.13.3	7.10
6.4	6.4	6.31, 8.20	9.13.4	9.13.4	7.10
7.1	7.1	6.20, 7.17	10.1	10.1	2.21, 6.31, 7.7, 7.13
7.2	7.2	6.22, 6.52	10.2	10.2	7.7
7.3	7.3	6.23	10.3	10.3	7.7, 7.14–7.15
7.4.1	7.4.1	6.52	10.4	10.4	7.16
7.4.2	7.4.2	6.53	10.5	10.5	7.15
7.5	7.5	6.23, 6.37	10.6	10.6	6.31, 7.15–7.16
7.5.1	7.5.1	6.29	11.1	11.1	2.42–2.43
7.5.3	7.5.3	6.30	11.4	11.4	2.31
7.5.5	7.5.5	6.36	11.7	11.7	2.38
7.5.6	7.5.6	6.28	11.8	11.8	3.40, 5.26
7.6	7.6	6.45, 6.52	11.9	11.9	3.40, 9.17
7.7	7.7	6.45, 7.22	12.1	12.1	9.6, 9.16
7.8	7.8	6.53	12.1.3	12.1.3	9.12
7.9	7.9	6.49	12.1.4	12.1.4	9.13
7.11.1	7.11.1	7.18	12.2	12.2	9.6, 9.16, 9.18
7.11.2	7.11.2	7.19	12.3	12.3	6.60, 12.19
7.11.3	7.11.3	7.19	12.4	12.4	9.19, 9.22
7.11.4	7.11.4	7.20	12.5	12.5	9.24
7.12	7.12	7.21	12.6	12.6	9.27
7.12.1	7.12.1	7.21	12.7	12.7	9.31
7.12.2	7.12.2	6.54, 7.21	12.7.1	12.7.1	9.32
7.13	7.13	6.54, 7.22	12.8	12.8	9.31–9.32
7.14	7.14	6.35	12.9	12.9	9.31, 9.33

Subject Index *by paragraph number*